THE FATE OF THE MAMMOTH

From Kolymsk to St. Petersburg. In the fall of 1901, a line of reindeer-drawn sleds of the St. Petersburg Academy of Sciences expedition crosses Siberia, carrying the remains of the Berezovka mammoth, carefully cut up, frozen, and wrapped in hides (V. E. Garutt personal collection.)

CLAUDINE COHEN

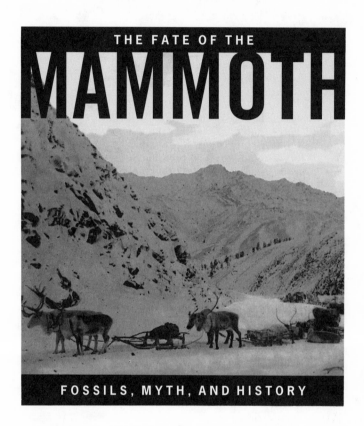

THE FATE OF THE
MAMMOTH

FOSSILS, MYTH, AND HISTORY

Translated by William Rodarmor

With a Foreword by Stephen Jay Gould

The University of Chicago Press
Chicago and London

CLAUDINE COHEN teaches the history of science at the
École des Hautes Études en Sciences Sociales (Paris)
and is the author of *Boucher de Perthes* (with Jean-
Jacques Hublin), *L'Homme des origines*, and *La Genèse de
Telliamed*.

WILLIAM RODARMOR is the translator of many French
books, including *Ultimate Game*, by Christian
Lehmann, and *Tamata and the Alliance*, by Bernard
Moitessier.

Originally published as *Le Destin du Mammouth*
© 1994 Editions du Seuil.

This work is published with the support of the French
Ministry of Culture—National Center for the Book.

The University of Chicago Press, Chicago 60637
The University of Chicago Press, Ltd., London
© 2002 by The University of Chicago
All rights reserved. Published 2002
Printed in the United States of America

11 10 09 08 07 06 05 04 03 02   1 2 3 4 5

ISBN: 0-226-11292-6 (cloth)

Library of Congress Cataloging-in-Publication Data

Cohen, Claudine.
    [Destin du mammouth. English]
    The fate of the mammoth : fossils, myths, and
history / Claudine Cohen; translated by William
Rodarmor; with a foreword by Stephen Jay Gould.
        p.   cm.
Includes bibliographical references and index.
ISBN 0-226-11292-6 (alk. paper)
    1. Mammoths. I. Title.
QE882.P8   C6413   2002
569′.67—dc21

                                                    2001048011

♾ The paper used in this publication meets the
minimum requirements of the American National
Standard for Information Sciences—Permanence
of Paper for Printed Library Materials, ANSI
Z39.48-1992.

POUR LOLA

CONTENTS

PART IV: SCENARIOS

ILLUSTRATIONS

As a professional and lifelong partisan, I have strong subjective reasons for regarding paleontology as the most fascinating subject in the universe. But this judgment can also claim some general, and hence objective, status—at least for entry into the top rank of favorites, if not for the absolute pinnacle— in the pervasity of public regard, as manifested across the full range, from pop culture's dinomania to intense debate among the "chattering classes" on patterns of human evolution or the moral implications (if any) of natural selection.

Much of the focus for this high regard lies in paleontology's intersection with two quite different paths of human fascination: our yearning to understand the timeless operation of the natural world by discovering the causes and principles of physical order; and our desire to document the specific and unpredictable pathways of life's actual history (the road truly followed among so many other potential, but unrealized, directions equally consistent with evolutionary theory). Paleontology makes a vital contribution to both these quests—to evolutionary theory, in the first category of general explanations, by providing our only source of direct data for the large-scale results of Darwinian (and other) processes in the fullness of geological time; and to tracing the pathways of history, in the second category of documentation, by studying the archives of bone and shell, the only remains of organisms that inhabited the earth so many million years ago.

(Despite their intimate associations, these two goals will always remain partly disassociated for a key reason beyond their different fascinations— general theory and actual history—for the human psyche; for no matter how precisely we may understand the principles of evolution, any general theory can only underpredict the actual pageant of life's realized history. Theory may dictate what cannot happen, but contingencies of particular moments— just as in the human history of rulers, battles, and nations—set the pathways actually followed among the innumerable possibilities permitted by nature's theoretical constraints.)

Given these dual sources of maximal interest, the question of how best to narrate the history of paleontological thought and action assumes particular importance. Most previous efforts have followed the time-honored convention of chronological accounting, as recorded by the lives and thoughts of the major scientists who strongly influenced the ever-changing history of our attitudes about the nature and meaning of fossils. But I have often thought that a reversal of this convention might yield even more insight—if only because this equally promising style of narration has been so rarely attempted. Why not relate the same chronology from the opposite perspective—that is, in terms of the organisms under scrutiny rather than the scientists doing the scrutinizing?

Accepting this premise of strong potential for a "reversed" history of the studied rather than the studier, which fossil organisms might be the best candidates for such treatments? Here we encounter a dilemma in trying to meet two criteria simultaneously: (1) length of debate, for fossils recognized only recently will not illustrate enough of the history of paleontological thought; and (2) degree of public interest and understanding evoked. Several candidates pass the first test of knowledge since the birth of the discipline, but fail the second standard of public concern—brachiopods, fossil corals, echinoderms, and many other invertebrates, with coiled ammonites perhaps asserting the closest claim to adequacy. Meanwhile, the two obvious standards for the second test of interest and empathy haven't been found or recognized as fossils until recently, and therefore do not tell a long enough tale of human investigation: dinosaurs, first defined in the 1840s, and humans themselves, not found as genuine fossils (despite scattered false claims) until the 1890s.

Among the few fossil groups that do qualify on both grounds, one candidate stands out as the obvious first choice: fossil elephants, with emphasis on the huge and extinct woolly mammoths. These creatures reside at the apex of fascination for the same set of reasons that led a famous childhood psychologist to the following crisp epitome for public (particularly children's) passion for dinosaurs: "big, fierce and extinct." (I'm not sure that woolly mammoths were truly fierce, but their large size and shaggy coats do inspire such an assumption, whereas their bigness, as terrestrial creatures, has only been exceeded by dinosaurs themselves.) Moreover, their extinction (in very recent times, after extensive interaction with early humans) lends exoticism both because absence itself makes the heart grow curious, while shaggy coats of cold-adapted species from times of ice ages seem so "unelephantlike" (by the parochial standards of modern survivors) that our interest grows apace.

Mammoths also surpass on the other criterion of maximal participation in the history of human thought about fossils. For mammoth bones (often confused with biblical giants in prescientific centuries) have stood out as prime subjects for extensive debate in every major episode of the history of paleontology. Mammoths have occupied public consciousness from the great Paleo-lithic cave painters of southern Europe, who drew mammoths from direct observation some ten to thirty-five thousand years ago; to modern molecular biologists who have extracted DNA from frozen tissues; to harebrained schemes in pop-culture versions, of bringing mammoths back to life by un-freezing some sperm, impregnating modern elephants, and after several gen-erations of inseminating the resulting offspring with revived sperm, breeding out the modern elephant almost entirely, while making an ever-purer mam-moth.

Some groping attempts to tell the history of paleontology through a mam-moth's eyes have been made before, but only as a lick and promise, and largely by amateur enthusiasts with (perhaps) adequate knowledge of the fossils, but little understanding of the subtleties or larger contexts in the history of sci-ence. But, in this truly pathbreaking book, the mammoth has finally met its match in Claudine Cohen, a distinguished French historian of science, who teaches in the capital of debate in Paris, where Buffon and Cuvier, from the mid-eighteenth to early nineteenth centuries, fueled the transition in our atti-tudes from mysterious, undated, isolated bones in gentlemen's "cabinets of curiosities" to entire reconstructed creatures with a long geological history of change and extinction.

We must resist the triumphalist interpretation of this chronology of ideas about mammoths as a steady and progressive march to ever-expanding and ever-perfecting knowledge, spurred by the "scientific method" and dispersing the previous darkness of theological and other forms of cultural bondage. Cohen, as a professional historian, knows in the deepest sense that such a cardboard accounting of human thought can only downplay and misrepresent the complex meanderings, social biases, and personal idiosyncrasies that make our unpredictable and ramifying histories so interesting and so informative as well. The history of science, perhaps more than any other subject, high-lights the human foibles and cognitive constraints that operate in constancy, as the actual content of belief shifts and grows for reasons spanning the gamut from changing whims of social fashion to genuine accumulation of new data. (In this sense, the history of ideas matches the paleontological pageant of life: both have been misinterpreted as predictable and linearly progressive; whereas both embody a primary theme of contingent ramification into a thousand byways of mental or physical possibility, while also, in a central and meaningful sense, adding something of value to an ever-branching structure.)

But, amidst all the branchings and dead ends, we can trace a chronology of accumulation and increasingly richer and more accurate interpretations. Any attempt to recognize objective temporal stages in this growing richness can only foster an endless wrangling that always accompanies the fool's errand of trying to break a true continuum into discrete parts. Still, the history of our growing understanding of mammoths in particular, and of the nature of fossils and the history of life in general, does describe a trajectory toward more accurate resolution—and this trajectory, as Cohen shows amidst her sensitivities to changing social contexts, does build a framework that converts this entire book into one exciting and continuous story. To avoid the fallacy of designating "stages" (as described above), we may, perhaps, best record this trajectory as a set of questions, answered sequentially:

1. What are these big (and spectacular) bonelike objects occasionally dug up in fields and then displayed in churches or in private collections? Are they truly the bones of creatures that once lived on earth? If so, what organisms do they represent? Are they the Nephilim described in Genesis 6?—"There were giants in the earth in those days." Or creatures killed by Noah's flood? Or remains of Hannibal's elephants from his ingenious, but unsuccessful, attack on Rome? Or perhaps they are "sports of nature," inorganic mimics of the mineral kingdom, and not the remains of organisms at all? As late as 1695, when the issue of the organic nature of fossils had largely been resolved in favor of our current view, the professors of the Medical College and Gotha proclaimed the famous Tonna fossil elephant to be an inorganic "imposter."

2. If we must admit that mammoth bones represent former organisms, and if they do not resemble any modern elephant in sufficient detail, could they represent species once vigorous, but now extinct? This commonsense solution of our day did not seem so obvious to scientists before the mideighteenth century—people who regarded the earth as but a few thousand years old, and who saw the organic world as a product of God's one-time creation in but a few days. After Buffon's documentation of the earth's considerable antiquity, followed by Cuvier's masterful work on the undeniable differences between mammoth bones and modern elephants (published in the last years of the eighteenth century), this revolutionary conclusion could no longer be doubted. But the implications and ramifications of this issue touch some of the great themes in the history of human understanding. American readers will particularly note Cohen's account of Thomas Jefferson's strong stand against extinction, his conviction that mammoths (if truly different from modern elephants) must still be living somewhere on the earth, and his partial rationale for sponsoring Lewis and Clark in a hope that these giant pachyderms might be hiding somewhere in the western United States.

3. If mammoths represent "lost" species, and the history of life therefore records substantial changes, what causes lie behind these alterations? This great issue, argued in starkest form between Cuvier's discovery in the late eighteenth century and Darwin's publication of the *Origin of Species* in 1859, initially pitted various forms of non-evolutionary thinking (from Cuvier's cat-

astrophic sequence of rapid extinctions followed by re-creations, to Lyellian views of continuous creation through time) against the conclusion that elephants (and all other creatures) evolve by a process that Darwin called "descent with modification." But the ultimate decisions for evolution did not resolve all debate by any means, for so many new questions arose, and so many old issues still required answers in the new evolutionary framework. What is the pattern of evolutionary change—a linear series of progress (and where, then, do mammoths stand in relation to modern elephants), or a branching tree with mammoths as a terminal bough and not as ancestors of modern forms? What causes evolutionary change—a process of inherent advance, perhaps theologically engendered or at least watched over, or a material process like Darwinian natural selection?

We have resolved these basic issues, but the story continues and new puzzles continually emerge as novel modes of study push our knowledge further. To end with a wry and specific tale, cited here as a symbol for viewing the explanation of mammoths as a work in vigorous progress, and not as a settled tale fit only for self-congratulatory narration, I cite the immortal words of my dear colleague, the late Allan Wilson, when he presented the first reliable DNA-based data on the relationship of mammoths to modern organisms. (To get the point of this story, one must know that DNA data frequently become contaminated with bacterial or mycological material): "We have learned today that the mammoth was either an elephant or a fungus." The beast is dead, but the beat goes on.

Stephen Jay Gould

Some seven years have passed since this book was first published in France, and I am happy to see that during that span of time it has achieved its goal of being read as dealing with both the study of mammoths and with the history of paleontology.

Important work in these two areas has produced new material, which has been included in this American edition.[1] Regarding the history of paleontology, several original texts of eminent scientists mentioned in the book have been edited or translated and are now available to the general public. Studies in the history of collections and museums have developed into a broad field of research, underscoring the importance of the history and classification of objects in shaping scientific knowledge. The philosophy and history of evolutionary theory have been the focus of new studies. A number of biographical monographs and general studies in the history of geology and paleontology have brought better insights into the intellectual, institutional, and sociological context of paleontological knowledge.

Scientific studies of the mammoth, in the field as well as in the laboratory, have been very productive, confirming that although mammoths have been the objects of intense scientific research for more than four centuries, they still have a great destiny ahead of them. Mammoth studies are a wonderfully active field of multidisciplinary and international research today, as they have been repeatedly in the past. The two first "International Mammoth Conferences," held in St. Petersburg in 1995 and in Rotterdam in 1999, brought together paleontologists, molecular biologists, archaeologists, specialists in Paleolithic art, and historians from all over the world to discuss their research and exchange knowledge about their totem animal.

Recently, DNA studies have yielded some unexpected and extraordinary results: laboratory research has concluded that mammoths may be genetically closer to African than to Indian elephants.[2] This conclusion, which is contrary to all that was previously believed, appears to be truly revolutionary. It must now be confirmed and reconciled with the conclusions based on anatomical studies.

In the field, research in Morocco led by a French team from Montpellier has discovered the oldest ancestor of the proboscideans: *Phosphatherium escuillei*, which was no bigger than a small dog, already had a trunk, and lived 59 million years ago.[3] But one of the most spectacular finds was that of "Jarkov," a frozen mammoth found in the Taymyr Peninsula in northern Siberia in 1997. Impressive pictures from that northern expedition have appeared all over the world.[4] In the beautiful light of the polar regions, these pictures show the animals, the landscapes, the faces of people, and, especially, the carcass of the frozen mammoth—still embedded in ice—flying suspended under a helicopter. They have popularized the romantic sense of the Arctic as one of the last frontiers and paleontological research as one of the last remaining adventures. Happily, this may also have spread the idea that mammoths are at least as interesting to search for and to study as dinosaurs.

Jarkov was not the first animal to be found with its flesh and hair intact, nor was it the first frozen mammoth attempted to be preserved whole. Exactly a hundred years ago, at the very beginning of the twentieth century, the St. Petersburg Academy of Sciences sent an extraordinary expedition led by a team of naturalists to unearth a whole mammoth found within the permafrost on the bank of the Berezovka River in Siberia. They attempted to keep the animal frozen, so as to make its soft parts—flesh, hair, skin, internal organs, and even digestive tract—available to scientific study.[5] A century later Jarkov's discoverers—a mixed team of amateurs and professional scientists—also decided to leave it frozen in its icy sarcophagus, in order to study it thoroughly, and maybe to clone it.

Indeed, after the achievement of cloning the sheep Dolly in 1997, research in fossil DNA—widely popularized by novels, movies, and the media—has nourished even more vivid dreams of cloning the mammoth. This would bring an extinct prehistoric animal back to life for the first time and make it possible to see herds of woolly mammoths again roaming the steppes of Siberia and Alaska. Would a "Pleistocene Park" be possible? In Japan, France, and other countries, some scientists have commented at length on what still remains today more a matter of fiction than of serious science. In June 1999 in the United States, a group of eminent paleontologists organized a memorial service at the Mammoth Site in South Dakota in June 1999 to mourn the extinction of the mammoth, and they expressed the dream of reviving the lost megafauna of the Ice Age by reintroducing their closest living relatives in the American West.[6] Indeed, fantasies about the mammoth are alive, and our woolly friend is more present today than ever in scientific discourse and in our dreams.

In recent years the mammoth has gained immense popularity in France, ever since a minister of education (who happens also to be a renowned geophysicist) chose it as the symbol for the heaviness of the French education system: "The mammoth has to be put on a diet," he claimed, to justify his reforms and budget cuts. Following that perilous proposition, hundreds of caricatures and cartoons were published in magazines and newspapers, showing the minister and the "mammoth" in various contexts and attitudes. The story ended three year later, when hundreds of thousands of students and professors, protesting the proposals, marched in the streets of Paris behind life-size portraits of mammoths, chanting *"Le mammouth est dans la rue"* (The mammoth is in the street) and eventually forcing the minister to resign.

Having this book translated into English has been quite an adventure, which, fortunately, has had a happy ending. Here I want to thank all the people who helped me in this matter: Martin Rudwick, James Secord, Niles Eldredge, Ronald Rainger, Roger Ariew, I. Bernard Cohen, Kevin Padian, Armand de Ricqlès, Jean Sudre, and Régis Debruyne, who kindly gave me an account of his latest and yet unpublished research on mammoth DNA. My work on this new edition has been supported by grants from the Canadian embassy in France, the Dibner Institute for the History of Science and Technology (Cambridge, Massachusetts), and the New York Public Library Center for Scholars and Writers. I owe many thanks to Jennie Dorny, Catherine Rouslin and Mireille Reissoulet at the Editions du Seuil, and to Susan Abrams and Erin DeWitt at the University of Chicago Press, whose patience and efficiency brought this project to a good end.

Last but not least, I want to express all my gratitude to William Rodarmor, who handled the *mammoth* task of translating the book with good humor, wit, elegance, and talent.

New York, October 2001

"On shells and on systems built on shells." Such is the title of a text in which Voltaire ironically wrote about his contemporaries who tried to explain the presence of marine fossils on mountaintops by constructing grand systems that would tell the history of the world from its very beginnings. Being a sensible man, Voltaire preferred to think that these shells and fishes were simply Roman kitchen scraps or the remains of meals eaten by pilgrims on their way to Compostela.

"On mammoths and on systems built on mammoth bones" could have been the title of this book. I cannot say exactly when my plan to write a general history of paleontology changed into that of a history of "systems built on mammoth bones." I think my careful reading over the years of the work of Cuvier, who founded scientific paleontology on the study of "fossil elephants," was decisive, as was that of an article by André Leroi-Gourhan called "The Mammoth in Mythic Eskimo Zoology."[1] "The history of the mammoth," that great prehistorian wrote in 1935, "is one of the most interesting chapters of the northern Pacific bestiary." It occurred to me that the history of the mammoth was also one of the most interesting chapters in the history of paleontology.

I started my research at the American Museum of Natural History in New York. I walked down galleries above which loomed the skeletons of dinosaurs and great mammoths to the Osborn Library, where I threw myself into reading old books and articles with accounts of excavation digs, all carefully dated and filed. I continued this exploration at the Bibliothèque Nationale and the Muséum National d'Histoire Naturelle in Paris, the British Museum in London, the Museum of Comparative Zoology at Harvard, and the Philosophical Society Library in Philadelphia. In the archives of the Academy of Sciences and the library of the Zoological Institute in St. Petersburg, I read the accounts of Russian expeditions to Siberia from the early eighteenth century through to the present.

I paid countless visits to the galleries of the Muséum National d'Histoire Naturelle in Paris and the Musée des Antiquités Nationales at Saint-

Germain-en-Laye. At the National Anthropology Museum in Brno, I saw the reconstructions of Paleolithic dwellings of Předmost and Dolni Vestonice in central Moravia; the animated mammoths at the Cardiff museum in Wales; and at the Mexico City museum, the nearly complete skeleton of a mammoth discovered not far from the Aztec city of Teotihuacán. I visited the Kostienki archaeological site near Voronezh on the Don River in Russia, where extraordinary dwellings made of mammoth bones have been uncovered. In Fairbanks, Alaska, I met a gold miner who spent every summer melting the permafrost with a jet of hot water to reach the gold-bearing quartz layer below. Advancing yard by yard through the hilly landscape, he turned the Pleistocene layers into a blackish mud that sometimes yielded mammoth bones, teeth, and tusks. In a store littered with tusks, I saw a carver using a dentist's drill to work ivory that had come down through the millennia. In an extraordinary shop in Anchorage—amid a jumble of whale vertebrae, Eskimo masks, and jade carvings—I saw mammoth teeth and ivory necklaces.

Gradually, this huge animal started to take over my life. I now have a vast collection of specialized paleontological and anthropological publications, images, and photographs. In addition to small carvings of mammoths in a variety of shapes and colors, I have collected more cumbersome specimens: a half jawbone with a superb molar brought up from the sea off the coast of Holland—a birthday present from my friend Peter Brinkmann; and an upper molar, a gift from the Kostienki archaeologists (It is somewhat crumbly, but the paleontologist Pascal Tassy, who consolidated it, tells me it's "an unusually large M3"). I also have mammoth hairs from Siberia and a pretty bracelet made of mammoth ivory.

So I want to pay homage to this venerable, cosmopolitan animal whose tracks have sent me crisscrossing almost the entire Northern Hemisphere. And I want to thank all those people who made this book possible, starting with Jean-Marc Lévy-Leblond and Isabelle Wagner, whose trust and patience were precious and of lasting help; Marie-Noëlle Bourguet, Jean Ehard, Jean Gayon, Charles C. Gillispie, Arno Mayer, and Pascal Tassy, who kindly agreed to read some of the chapters. Mikhail Anikovitch, Michel Blay, Eric Buffetaut, Yves Coppens, Roger Hahn, John Heilbron, Frederick L. Holmes, Ian Jelinek, Adrian Lister, Dick Moll, Michael Novacek, Martin Oliva, Nikolai D. Praslov, Ronald Rainger, Richard Tedford, Hans Van Essen, Alan Mann and Michèle Lempel, and Dale and Marylee Guthrie, who all helped me with my research. I owe a great deal to my friends Paul Bahn, Sylvie Bleuse, Peter Brinkmann, Anne Delmer, Michel Garcia, Cyrille Galpérin, Kevin Matthews, and Jean and Floria Prodomidès. To Vadim E. Garutt and Nikolai Vereshchagin of the St. Petersburg Zoological Institute, and Roček Zbyněk of the Geological Laboratory of the Prague Academy of Sciences, for the priceless gift of lending me rare reproductions and photographs. To Stephen Jay Gould, who several times welcomed me with the greatest human and intellectual generosity to the Museum of Comparative Zoology at Harvard, and who gave me permission to copy a precious old collection bearing Blumenbach's handwritten notes. And to Natalie Zemon Davis, thanks to whom I was able to finish the writing of this book in the warm and studious atmosphere of the Davis Center at Princeton, albeit amid the blizzards of a quasi-Siberian winter.

This book has been partly funded by the Centre National des Lettres, the Centre National de la Recherche Scientifique, the Commission Franco-Américaine, the Fulbright Foundation, and the Shelby Cullom Davis Center for Historical Studies of Princeton University. The École des hautes études en sciences sociales, to which I belong, has given me the means of conducting my research under good conditions. The students and colleagues who participated in my seminars at the Centre Alexandre-Koyré stimulated my thinking with their questions and remarks, and I extend them my warmest thanks.

Paris, April 1994

This is not a book about mammoths.

The story you are about to read is not the grand, tragic tale of a large mammal with heavy curved tusks and a massive body covered with long brown hair, which appeared about 400,000 years ago and disappeared some 10,000 years before the present, and maybe even more recently.

Here, the mammoth is only a pretext, a basis for this book. What is really involved is the history of paleontology—the history of systems of interpretation built on fossil objects. It is the study of how, for more than three centuries, those objects have been part of legends and fables, of narratives of the history of the earth and of the evolution of life. For the first time, a history of paleontology will follow changing perspectives on a single object: the enormous remains—teeth, bones, and frozen flesh—of what the native Siberians called in their language *"mammut"* and what Western scientists, in theirs, have named "figured stones," "bones of giants," "fossil unicorn," *Elephas primigenius*, and *Mammuthus meridionalis, columbi, imperator.* In the course of the history of knowledge, those terms have referred to various ways of framing discourse and to different systems of thought and interpretation. What is at stake here is the meaning these strange remains from the earth have taken on in various times and places, in human life and culture.

The history of paleontology has often been told as the history of ideas,[1] as the progressive growth of a body of knowledge that at the end of the Renaissance freed itself from medieval notions of "sports of nature" or "figured stones" and finally determined that fossil objects were the remains of living beings. These debates, which occurred throughout Europe between the sixteenth and the eighteenth centuries, gave rise to the construction of "theories of the earth," in which fossils were often believed to be "medals of the Deluge," remains of the biblical Flood. In Georges Cuvier's work in early-nineteenth-century France, the recognition of fossil vertebrates as extinct animals opened the dizzying perspective of a succession of "lost worlds" destroyed by terrible cataclysms. A few decades later, Darwin's theory in *On the Origin of Species by Means of Natural Selection* proposed the idea of an evolution that linked all living beings from their single common origin to the appearance of present-day plants, animals, and humans. This renewed the concepts and methods of paleontological research, the goal of which became to discover the missing links in the history of life.

But this picture of a science gradually shedding naive beliefs in favor of rational knowledge needs to be reexamined. More detailed studies have yielded a complex tableau composed of traditions and controversies, eclipses and reinterpretations, intellectual revolutions, and, at times, returns to theories once thought obsolete.[2] Moreover, the history of paleontology is not only a history of scientific ideas. It is deeply tied to the history of scientific practices—the collecting of buried bones, the organization of digs and expeditions, the meticulous cleaning of fossil objects and their description and classification in collections. It is tied as well to representations of the world and changing social and cultural configurations within which these stories make sense.

The history of paleontology has also been told as the story of the heroes of science. Since Cuvier, an enduring myth exists of the paleontologist as a new hero, "*un antiquaire d'une nouvelle espèce*," who is able to decipher the vanished monuments of the past, to dig them out of the layers of the earth and spectacularly reconstruct monstrous animals from small pieces of bones, thus giving new life to entire extinct faunas. Making the history of science a gallery of great names—in this case the names of Cuvier, d'Orbigny, Gaudry, Boucher de Perthes, and Boule in France; Darwin, Falconer, Owen, and Huxley in England; Leidy, Marsh, Cope, and Agassiz in the United States—would follow this heroic trend. But it would not fully illuminate the relationship of knowledge to its ideological, institutional, sociological, political, and economic context. Nor would it help us really understand the transformations of the images and stories by which this knowledge has been developed, shared, and popularized.

This book aims to tell the history of paleontology not just as a history of discoveries, ideas, or practitioners of science, but through the *history of objects*, of these strange objects—teeth, bones, and shells—called "fossils" that have been dug up in excavations and expeditions to the ends of the world, extracted from the layers of the earth, conserved and gathered through specific techniques, and arranged in museum collections or evolution galleries. Pride of place has been given here to a single "object" in order to shed light on the transformations of paleontological methods and knowledge, an object where myths, questions, and stories converge, an object with the dimensions and the density of an enormous "thing"—mammoth bones that are dug up, reassembled, studied and admired—which is also the constructed object of scientific practices, discourses, and theories.

In reconstructing extinct species, paleontology asks questions about the deep past of the earth, about the origins and evolution of living organisms and specifically of humans. These questions inevitably encounter those of myths. For a long time, the discourse on the origin of fossils and the transformation of the earth and life borrowed the framework of the biblical narrative, with its miraculous causes and a time frame limited to some six thousand years. The result was the story of the successive appearance of living beings, from plants to marine animals to land animals, of which humans were the triumphant end result. It assumed the image of a living world unchanged since its very beginnings, its inhabitants today being just the way they were on the day of Creation. The history of the earth included such standard episodes as the Flood, a universal cataclysm seen as a major event in the history of the world. These themes and narrative schemes survived well beyond the moment when paleontology became a scientific discipline at the start of the nineteenth century.

Paleontology borrows from myth, but it also borrows from history. Paleontologists, like historians, collect fossilized "documents" from the "archives of the earth." These are partial, scattered, fragmentary objects, but they are unique, irreplaceable witnesses to the past. Like the historian, the paleontologist uses "documents" from the past to weave a story, spinning a plausible account in which facts are arranged along the axis of time. Shaping these models of time and change, of events, causes, and actors, necessarily leads to the elaboration of a narrative. A narrative builds an explanation, organizing scattered vestiges into a story that gives them meaning and coherence. Naturally, the goal is to reconstitute the one true history of the earth and of life. But there are several possible ways of telling that story, depending on the facts given, the conceptions of time, and the supposed tempo and modes of evolution.

Naturalists dealing with "deep time" have often borrowed from historians the ways in which they framed their discourse. Georges-Louis Leclerc de Buffon thought of epochs of nature the way Jacques-Bénigne Bossuet thought of epochs of human history. Cuvier's *Révolutions de la surface du globe* borrowed from the grand history of the late eighteenth century and the historiography of the French Revolution. And throughout the nineteenth century, the concept of a progressive evolution of living beings that reached its height with the appearance of man was modeled on the great historical frescoes depicting the progress of humanity.

In the mid-nineteenth century, the paleontological history of species took the form of a genealogy. The evolution that Darwin proposed to reconstruct was no longer determined by a divine plan or by a finality imposed from outside or by the theological scheme of a progression leading to man. The evolution of living beings, the appearance of species and their survival and extinction, was the result of their capacity to adapt to their environment, of their struggles to survive. From then on, paleontological research began to focus on the search for "missing links," and especially on the most spectacular phases of evolution: the origin of plants and animals; the first vertebrates; the origin of terrestrial animals, mammals, and birds; and the origin of humans.

And yet when it became a matter of reconstituting "the origin of species by means of natural selection," the ambition of encompassing the whole history of the living world was out-of-date. Darwin himself regretted the gaps in the fossil record that made it impossible to reconstruct the total fabric of the evolution of life. For a number of paleontologists today, the whole history of life is not necessarily one pointing toward inevitable progress.[3] The global rationality of the history of the living world has exploded into a multiplicity of events that cannot all be taken in with a single glance. It becomes important to consider not only "laws" of biological evolution, but also accidents of environment and the role of contingency in shaping the evolution of life. To the paleontologist now falls the task of reconstructing not vast frescoes or all-encompassing narratives, but more modest biological, geological, or ecological "scenarios" that come into play at particular events in the history of living organisms: the extraordinary evolutionary flowering of invertebrates in the Cambrian era; the extinction of the dinosaurs at the end of the Secondary; the diversification of species of fossil mammals in the Tertiary; the coexistence of several species of hominids during the Pleistocene; and also the details of evolutionary history written in the morphology of every living or fossil creature: the "panda's thumb,"[4] the structure of a horse's hoof, or the strange shape of an elephant's molar.

To a modern paleontologist, "scenarios are inductive narratives . . . concocted to explain how some particular configuration of [evolutionary] events . . . took place," writes Niles Eldredge. "Such narratives . . . concern themselves with the application of what we think we know about evolution to the real world, as preserved in the fossil record. But they are mainly fairy tales constructed of a maze of untestable propositions concerning selection, function, niche utilization, and community integration." Scenarios are speculative constructs, and they are sometimes risky ("fairy tales"). Nevertheless, they are valuable and useful to science as heuristic devices, and they force us to "stretch our imagination" to invent new hypotheses. Eldredge explains: "As long as we explicitly realize how we build scenarios and what their status is as scientific propositions, we should continue to build them, hoping to find them scientifically, as well as spiritually, uplifting and rewarding."[5]

Myths, stories, and scenarios not only represent three aspects of paleontological discourse, but also the dominant forms of this discourse at different moments of its history. In presenting hypotheses in the form of "scenarios," paleontology has not left the realm of history. It continues to rely on the representation of an immense temporality, an irreversible evolution of beings from which laws can be derived. And in the way it shapes its themes and narratives, paleontology remains secretly inhabited by myths. As a quest for something other and elsewhere, paleontological research includes the dimension of the heterogeneous, the strange, the enormous. It achieves the dream of traveling through time, visiting the depths of a vanished past that is terrifying but harmless, since it takes the form of desiccated remains of extinct monsters that appeal to our imagination. These fascinating or grotesque animals, these strange events and cataclysms that inhabit paleontologists' discourse also take on a wholly imaginary configuration: some of the themes recur periodically or live on in scientific formulations. And whether the element of myth and fiction is related to the canonical shapes of the stories that inform our culture or to the very structure of our imagination, it is present in paleontology both in scientific discourse and in its popularization.

The mammoth was chosen as the thread running through this book. Other choices would certainly have been possible. Different fossil species have taken center stage at different times, depending on the questions being asked. Throughout the Renaissance and until the end of the eighteenth century, seashells and fossil fishes discovered underground and even on mountaintops fascinated naturalists and lovers of curiosities, and sparked heated arguments about how those marine remains wound up in the mountains. From the end of the eighteenth century, the focus shifted to the bones of quadrupeds, leading to the debate over extinct species and the history of the living world. In the mid-nineteenth century, the search for fossil hominids became the primary focus of attention and research, raising new questions about man's place in nature. But interest in fossil invertebrates did not disappear when attention shifted to quadruped bones or questions about human fossils, and pulling on all those threads at once would have created a very tangled skein.

The mammoth has the advantage of being a massive, continuous presence in Western knowledge from the end of the Renaissance until today. When the Dutch traveler Nicolaas Witsen first named the Siberian natives' "*mammont* bones" in his *North- and East Tartary* in 1692, Europeans did not grasp the fact that they belonged to an extinct animal. Until late in the seventeenth century, a good number of those who studied fossil bones believed they were produced by the earth itself, as "sports of nature," "figured stones," or fragments of living beings born of seeds carried beneath mountains by underground canals. When travelers reported the beliefs of the inhabitants of the far north, Europeans wondered whether these could be the remains of giants or unicorns, whales or elephants. Until the last decades of the eighteenth century in Germany, France, England, and the United States, sober debates were held over how those elephants could have reached the frozen north of Asia or America. Had they been carried there by the waters of the Flood? Were they African elephants brought by Hannibal's armies who had escaped to northern climes? Or could they be the bones of animals still living in unexplored parts of the earth?

Countering these "fables," Cuvier in 1796 masterfully demonstrated that the remains were in fact those of a "lost" species. In his *Mémoire sur les espèces d'éléphans vivantes et fossiles,* he argued that the mammoth had become extinct and suggested a history of the earth in which all living beings had periodically succumbed to gigantic cataclysms. Cuvier was henceforth hailed as the founder of a new discipline that would take its place in learned institutions—with academic and university chairs, museums, and periodicals—and would come to rely on an international network of amateur and professional scientists.

But in founding paleontology—the study of extinct beings—Cuvier devised a system of thought (geological catastrophism and its corollary, biological fixism) that also was a fiction, a new myth. In his account of the history of the earth, he used powerful images likely to stir his contemporaries' imaginations, evoking the violent cataclysms of "the revolutions of the surface of the globe," the gigantic "pachyderms" that once populated the earth, and the compelling image of a science that could bring vanished worlds back to life.

The mammoth appears as an emblematic figure in paleontology not only because it was the first extinct animal to be identified and reconstructed. It also belongs to an order so diverse and rich, that it is a special focus of the classification theories that are a major issue of contemporary paleontological research. Cuvier defined a single species of fossil elephants and the extinct genus Mastodon, which he thought was represented by "one or two species," but knowledge about the proboscideans has increased dramatically since then. The richness of this once widespread group is barely hinted at by the two (or maybe three) species of elephants found today in India and Africa. Mastodons belong to a very diversified family of primitive proboscideans that appeared at the beginning of the Miocene some 20 million years ago and became extinct about ten thousand years ago. Cuvier's three species of the genus *Elephas* have become three distinct genera with a common origin but separate histories: *Loxodon, Elephas,* and *Mammuthus* constitute the family Elephantidae. The woolly mammoth is one of the branches, today extinct, of the genus *Mammuthus.* This taxonomic term, which was first used by Joshua Brookes in 1828, has only really been in scientific usage since 1935 to designate a genus with its own history, independent of that of the other Elephantidae. The mammoth is nowadays involved in many key questions of paleontological research. The great diversification and expansion of the genus throughout the Northern Hemisphere, from the middle of the Pliocene (3 to 4 million years ago) to the end of the Pleistocene, made it necessary to construct scenarios of evolution and migration. As with dinosaurs, the extinction of the mammoths remains an open question. And frozen specimens have been found in such good condition that molecular biologists can extract DNA and attempt research on it.

The mammoth is the totem animal of vertebrate paleontology, but it plays the same role in human prehistory. It is the symbol of a vanished but familiar era, a symbol of those ice ages that in our mind tend to merge with the earliest history of humanity. The mammoth was the contemporary of the great prehistoric hunters, the "cavemen." One could also say that the mammoth was tied to the discovery of human fossils, in two ways. When mammoth bones were found in conjunction with man-made flint objects, they proved that humans existed at the same time as "antediluvian" animals. And when Edouard Lartet in 1864 discovered a piece of mammoth ivory inscribed with an image of the animal itself, it confirmed not only the great antiquity of hominids, but also those "primitive" people's artistic ability.

The mammoth is proof of the diffusion of traditions and knowledge. It is hard to ignore its remains, since mammoth bones, teeth, and tusks can be found from Italy to China and Siberia and from the depths of Alaska to southern Mexico. Their presence is so obvious, it forces us to have an opinion about them, to tell a story that legitimizes their existence, their presence. The mammoth also bears witness to the coexistence of several types of representations in different places at the same time, as well as of the national particularities of the various "styles" and scientific schools. Russian, English, German, French, and American paleontology and prehistory have different methods and research traditions, because each takes place within a particular geological, intellectual, and institutional context.

The mammoth may be the best known of fossil animals, but that does not prevent it from being the object of projections, representations, and extraordinarily varied dreams—quite the contrary. Today the mammoth is the object of serious study by paleontologists, systematists, and geneticists, but it is also a character in novels and comic strips, a star of movies and advertising. An array of views, stories, and images are connected to this big animal, which, along with the dinosaur, is one of the most vivid heroes in our paleontological imagination.

The mammoth is not only the object of abstract speculations, but also the object of material use and commercial exchanges. Fossils have usefulness and trade value in addition to their scientific worth. From the ninth century on, fossil mammoth ivory was a trade item between Chinese and Arabs. Today this extinct and precious pachyderm has replaced the African elephant as a source of ivory, thereby coming to the rescue of its cousin, which is itself now threatened with extinction.

So the fate of this volume is linked to that of the mammoth, but the book also aims to be a meditation on the history of scientific knowledge. I would like the reader to see a double allegory in this venerable and hairy animal: on the one hand, the crystallization of problems that arise from the study of science, as it attempts to articulate the history of ideas, institutions, practices, and discourses; on the other, the value of considering the specificity of a discipline that stands at the crossroad of the sciences of nature and of history. What is involved here is the way in which fossil objects—through these successive representations, images, and stories—have given rise to many systems of interpretation linked to changing social and cultural configurations. Once again, the mammoth will be the basis of a story—that of the quest for the prehistoric past of life.

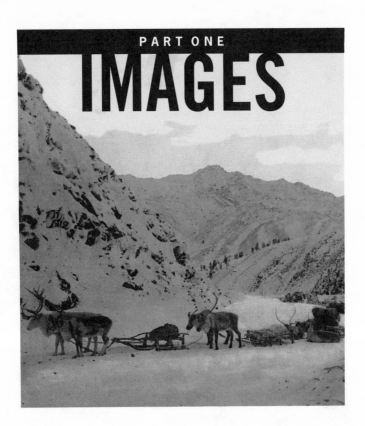

PART ONE

IMAGES

La Fuite devant un mammouth (Fleeing from a mammoth). Gouache by Paul Jamin (1906). (Musée des Antiquités Nationales, Saint-Germain-en-Laye.)

The Mammoth Appears

La Fuite devant un mammouth (Fleeing from a mammoth) is the title of a painting you can see at the Musée des Antiquités Nationales at Saint-Germain-en-Laye, near Paris. The 1906 work is by a somewhat forgotten French painter named Paul Jamin. Its right-hand edge bears a clearly legible dedication to Louis Capitan, one of the great prehistorians of the time.

In the painting, a mammoth stands in the background, its trunk raised, looming heavy and dark against the frozen landscape. A few stunted trees rise from a snow-covered hill. In the foreground, four half-naked men are scrambling pell-mell, their arms flailing, fleeing the colossus. The scene, which is intended to show our ancestors' lives in a hostile environment, emanates a sense of desolation and terror.

Of strangeness, too.

The mammoth's tusks curve outward, as if backward, and the men look like cartoon savages, with their animal skin rags, claw necklaces, and puny weapons—chipped flint ax-heads lashed to wooden handles. But the painting, in bringing together in the same frozen landscape a group of caricature primitives and a naturalized specimen straight out of some natural history museum's paleontology hall, sums up the way prehistory was imagined at the time, with all the anthropological, paleontological, and geological clichés of the late nineteenth century. The mammoth, with its outspread tusks, is the very image of a mounted specimen from the Lena River displayed in the St. Petersburg Zoological Institute until the end of the century. The vaguely Mongol "savages" embody the "primitive man" so dear to anthropologists of the period. And the frozen wastes suggest the indefinite "glacial period" that geologists then conceived it.

In blending imagination and the state of knowledge of the time, the painting embodies an outmoded vision of the earth's prehistoric past, of animals, and of man. It shows that paleontological reconstructions have their own history, too.

When an entire frozen mammoth carcass was found at the mouth of the Lena River in Siberia in 1799, its tusks had been cut off and sold in Yakutsk by the Tungus man who discovered it, and scavengers had already gnawed at the carcass. The botanist Mikhail Ivanovich Adams tried to bring it back to St. Petersburg in 1806 but was only able to return with the nearly complete skeleton and some soft tissue, including the head with an eye and an ear, and the lower parts of two feet, which had been preserved in the frozen ground.[1]

In images then current in Russia, the mammoth was thought to be an animal without a trunk, but with two long, downward-pointing "canines." Or else it was a kind of boar with a wrinkled hide and tusks that flared out like a mustache. Other engravings show an animal with braided horns high on its forehead, like the single horn of the unicorn.

The Lena River specimen was the first complete mammoth skeleton ever to be reconstructed. It was initially displayed in Peter the Great's *Kunstkammer*, or cabinet of curiosities, then transferred to what is now the St. Petersburg Zoological Institute, where it is today (see photo on page 113). But the tusks thought to belong to the animal had been mounted backward, with their tips pointing outward. In 1899 they were moved to the more accurate position with their tips curving inward.

An anatomical reconstruction in paleontology has often been presented as an infallible operation. The paleontologist is thought to have an almost magical power to bring beings from the past to life, as if by the sound of the Judgment Day trumpet. In an enduring myth, scientists take a fragment of bone or a single tooth and reconstruct the entire skeleton of a huge dinosaur. In fact, reconstruction is a complex operation that combines thought and imagination with practical skill and a gift for comparing objects. In some ways, it is like putting a puzzle together or assembling a mosaic. And reconstruction embodies not only science's concepts and techniques, but the images and dreams of an era. It renews its images and themes in response to empirical discoveries, but also to changes in the framework of knowledge.

In the late eighteenth century, Georges Cuvier identified the mammoth as belonging to an extinct species closely related to today's elephants. What was then known were its impressive size, coat of long reddish brown hair, dome-shaped skull, molars, and huge, curved, spiral tusks. But some unknowns remained, notably the position of the tusks, the skeletal structure, the outline of the back, the body's length-to-height ratio, and the arrangement of the soft tissues. In the mid-nineteenth century, Benjamin Waterhouse

This drawing of a mammoth—a bizarre cross between a bull and a unicorn, with its long tail, braided "horns" on its forehead, and claws—is attributed to a Swedish soldier taken prisoner by the Russians who crossed Siberia in 1722 on his way home to Sweden. Josef Augusta and Zdenek Burian, *A Book of Mammoths,* 1963. (Swedish library archive photo.)

Adams's mammoth, as drawn by the merchant Boltounov in 1804: an animal without a trunk that looks somewhat like a boar, with "canines" pointing down and out. (V. E. Garutt personal collection.)

Hawkins pictured the mammoth swaddled in fat, long-legged, and nearly hairless, with a bald trunk and a little mane behind its head.[2]

Throughout the nineteenth and twentieth centuries, a series of discoveries led to changes in that image.[3] Russian expeditions brought entire mammoths to light, with their flesh and bones preserved in ice, giving us—in a way unique in the world of fossils—the intact bodies of animals that disappeared from the surface of the earth thousands of years ago.[4] They provided better knowledge of the details of mammoth anatomy: the arrangement of the skeleton, the size of the animal's tusks, and the structure of its teeth—even the length of its hairs, its woolly coat, and the shape of its tail. The Siberian permafrost preserved the soft tissues, which normally do not fossilize, thereby revealing the thickness of the mammoth's subcutaneous fat, the contents of its stomach, even the length of its erect penis (three and a half feet). We now

With its wrinkled hide and tropical setting, this curious mammoth drawn by a Russian amateur bears little resemblance to the reconstructions that illustrated scientific works of the time. From an article by R. F. Tepkerp. (V. E. Garutt personal collection.)

The mammoth in the mid-nineteenth century: a long-legged, almost hairless elephant, larded with fat. Wall poster circa 1862 by Benjamin Waterhouse Hawkins for Britain's Department of Science and Art. (This illustration also appears in Martin Rudwick, *Scenes from Deep Time*, 1992, p. 165.)

know that the mammoth had hairs on its trunk and that its feet had four toes, where other Proboscidea have five. The anatomy and development of the young are better known, thanks to the 1977 discovery of a juvenile mammoth in Magadan, Siberia. It also seems that the size of mammoths varied with time and place. The oldest specimens in central and western Europe (the Siegfried mammoth, for example) stood as high as twelve feet at the shoulder and weighed six tons. But the more recent Siberian mammoths were smaller than Asian elephants (nine feet at the shoulder and "only" four tons). Still, myths about the mammoth's gigantic size persist. In our collective imagination, it remains a "monster" of prehistory, a theme inherited from ancient myths and expanded on in the nineteenth century, in particular in Cuvier's scientific work.

Since they were discovered in the early twentieth century, the pictures left by Paleolithic hunters—carved and drawn or painted on the walls of their caves—have changed the way the mammoth is represented. Some prehistori-

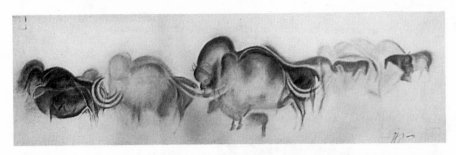

The mammoth in Paleolithic cave art: a frieze of bison and mammoths at Font-de-Gaume (Dordogne). Watercolor by abbé Henri Breuil. (Musée de l'Homme photo.)

ans have tried to deduce the animal's "real" appearance and behavior from the images in the painted caves of southwest France—La Mouthe, Font-de-Gaume, Les Combarelles, and Rouffignac.[5] But like all works of art, these images both represent and transform reality and are stylized to the point of caricature, with an occasional touch of humor. The prehistoric sculptor or painter accentuated some of the mammoth's peculiarities, sometimes reducing them to a few characteristic traits. In a painted or carved image

Various mammoth styles in Paleolithic art. *Left:* Painted "geometrical" mammoth in the Kapova Cave (Urals). (Nina Garutt gouache and photo.) *Right:* Engraved mammoth at Font-de-Gaume (Dordogne). (Abbé Henri Breuil drawing.)

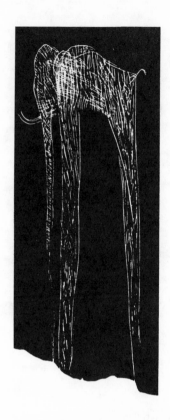

The Berelyokh (eastern Siberia) mammoth, inscribed on ivory. (N. K. Vereshchagin photo.)

of a mammoth, usually shown in profile, the artist almost always portrays its domed skull, like a bishop's biretta; nasal bones extending the forehead vertically; the compact, almost spherical body; the short, massive legs; the long hairs forming a "skirt" around the lower body; and the curve of the trunk and tusks.[6]

Images of mammoths are relatively rare in Western prehistoric art, except in certain unusual sites, like the Rouffignac Cave in Dordogne. But they are very common in eastern Europe, in Czechoslovakia, Ukraine, and Siberia, and follow highly stylized patterns. "Despite certain differences in execution," writes Zoia Abramova, "they all share common traits that reflect the mammoth's characteristic shape: a small, domed head, a large humped back, and a trunk that hangs well clear of the body. Almost all the figurines have a flat, often polished base, so they could be stood upright."[7]

The artists in these regions pushed originality and stylization very far. The "geometrical" mammoth from the Kapova Cave in the Urals could be the

Appearance of Man, revised version, from the sixth edition of Louis Figuier's *Earth before the Deluge* (1867). The presence of the mammoth attests to the "antediluvian" character of human existence.

work of a prehistoric Picasso. And what was the artist trying to express who carved, on a piece of ivory found at Berelyokh, a mammoth whose endless legs stretch down toward some unknown abyss?[8]

Inspired by these discoveries and their own imagination, engravers and painters of the second half of the nineteenth century tried to reconstruct not only the skeletal structure of animals from the past, but also their musculature, soft tissues, mode of locomotion, setting, and environment. This very special genre of animal art consists in bringing to life through images the physiognomy and attitudes of these beings from another age, and placing them in the context of their relations with other animals and with humans. Paleontology may be the only field where art forms an integral part of scientific work.

In the nineteenth century, painting "scenes from deep time" became a distinct pictorial genre, often inspired by biblical themes.[9] Thus, the first edition of Louis Figuier's *La Terre avant le Déluge* (The Earth before the Deluge) (1861), which popularized the geological and paleontological knowledge of his day, shows human beings in the quasi-biblical setting of the Garden of Eden. In 1867, when the existence of fossil man was established and accepted by the scientific community, Figuier published a new edition of his book, in which he modified the scene.[10] The "prehistoric" men, now outfitted with animal skins and primitive axes, mingle (at a safe distance) with "antediluvian" animals, including the obligatory mammoth. Henceforth, the mammoth would be part of the imagery of human prehistory. These engravings, which illustrated best-selling books and appeared in many editions, played an important role in representing prehistory at the time—and perhaps still do today.

In France at the end of the nineteenth century, "prehistoric painting" was a flourishing genre, and its works had a special status, at the intersection of scientific and popular knowledge. They summed up the representations of prehistory, biblical myths, and nationalistic themes related to France's "ancestors," the Gauls. Academic painters tried their hand at it, and the resulting masterpieces of pompous art were displayed in Paris at the Musée d'Orsay and the paleontology amphitheater in the Muséum National d'Histoire Naturelle. For example, the paintings of Ferdinand Piestre, who was known as Cormon,[11] are full of clichés about "primitive" peoples, caveman imagery, and biblical references.

Charles Knight painting in his Bronxville studio around 1911.
(American Museum of Natural History photo.)

Detail of a Charles Knight mural, *Reindeer and Mammoths on the Somme River, France, in Winter* (1916). (American Museum of Natural History photo.)

In the United States, paintings of prehistory became a scientific genre in its own right after the turn of the century. The paleontologist Henry Fairfield Osborn, who founded and ran the paleontology department at the American Museum of Natural History in New York until his death in 1936, used Charles Knight as his personal artist,[12] and Knight painted the paleontological and prehistoric murals that for many years graced the walls of the paleontology galleries. While the dinosaur may be the undisputed hero of American paleontological imagery, the mammoth runs a close second. For the museum's Age of Man gallery, the artist in 1916 painted a majestic panorama of the Pleistocene Ice Age that shows a herd of mammoths—massive, powerful, and monumental.

In 1920 Knight traveled to France to visit abbé Henri Breuil, then the reigning authority on prehistoric art, and to tour the painted caves of the Vézère Valley. For the American animal painter, seeing the Paleolithic works must have been astonishing. In the cave frescoes, the twentieth-century artist could see the reflection and model of his own activity. In a 1920 painting, Knight shows the Font-de-Gaume artists painting a frieze of mammoths and bison on the cave walls. Nearly naked in their loincloths, they are mixing ocher and charcoal colors on a reindeer scapula by the faint light of an animal-fat lamp.[13]

In 1940, as France was collapsing, four schoolboys would reveal to the world the art of the Lascaux Cave. In the following years, digs at Paleolithic sites in eastern Europe would generate fresh pictorial reconstitutions of prehistory. In Czechoslovakia between 1940 and 1960, the painter Zdenek Burian[14] collaborated with paleontologist Josef Augusta to produce illustrated books for the general public that would circulate throughout the world. *A Book of Mammoths*, published in Prague in 1962,[15] was the first large popular monograph devoted to the mammoth, and its illustrations were endlessly reproduced in textbooks and exhibit catalogs. Burian's reenactments reflect theories about mammoths and their way of life, how they were hunted and butchered, and how they became extinct. His images often focus on human activities connected to hunting, toolmaking, or art. The paintings are somewhat conventional in execution, but their focus on humans makes them touching. They show how hard life was in prehistoric times, representing it less as a mythology of conquest than as a struggle to survive.

Because of their narrative power, images tell many things about beings of past times. But words remain the highest form of expression of prehistoric fiction—storytelling, poetry, and literary creation. In a number of short stories and novels, man and mammoth come together. This "prehistory fiction"—a limited but very real body of work—has its own history and readership. Set in prehistoric times, the works bring descriptive or quantified scientific knowledge alive through "eyewitness" accounts, promoting a discourse that science is unable to achieve, while proving the popularity of the

Mammoths in the Snow, by Zdenek Burian (1961). (Moravské Zemské Muzeum–Anthropos, Brno.)

discipline. In France in the mid-nineteenth century, Elie Berthet and Félix Boitard brought Cuvier's "lost worlds" to life. Novels became didactic, a vehicle for theories and hypotheses. By the end of the century, a huge "popular" literature made the themes of paleontology and prehistory available to a large readership.[16] At the same time, a number of works of fiction were produced that remain well-known today. In France the first "prehistory novels" began to proliferate, written by the novelist Jules Verne[17] and the poet Edmond Haraucourt.[18] Prehistorians themselves presented their own discoveries in novelistic form, combining educational intent with the lure of storytelling. Often handsomely illustrated with engravings, the books may have been the best way to present and enliven the theories. The prehistorian Adrien Arcelin, one of the investigators of the Solutré site, used a pseudonym— Adrien Cranile—when he published a novel in 1872 called *Solutré, ou Les Chasseurs de rennes* (Solutré, or the reindeer hunters).[19] Through the magic of a dream, the book's hero finds himself carried backward into prehistory, which comes to life before his eyes. Romantic adventures ensue, involving a love affair between the hero and a chief's daughter and the obligatory mammoth hunt.

But the most famous "prehistory novels" were written at the turn of the twentieth century. In England H. G. Wells published his *Stories of the Stone Age* in 1900, while J. H. Rosny Aîné was publishing his famous prehistoric novels at about the same time in France.[20] *Vamireh*, Rosny Aîné's first such tale, came out in 1892, and *La Guerre du feu* (translated as *Quest for Fire*), which appeared in 1906, remains a model of the genre. Its story, which is set "a hundred thousand years ago," tries to bring to life the hypotheses and facts that were known at the time: paleontological (prehistoric animals, geography, climates, the Ice Age), paleoanthropological (pre-*sapiens* people encounter various ape-men), and archaeological (cannibalism, mastery of fire, material cultures). Narrative devices, not to mention clichés, occasionally poke through this "scientific" veneer, bolstering the story's effectiveness.

Some are characteristic of the narrative scheme of fables: for example, a binary split between the characters (some are good, handsome, and strong; others, nasty, cruel, and bestial—e.g., the Aghoo brothers) and the ternary structure of the episodes. The *Quest for Fire* plot resembles a fairy tale in its structure (bringing back a proof of one's courage to obtain the beloved woman) and also in its characters, who are arrayed in a Manichaean duality: the "good guys" are graceful and courageous; the "bad guys" are hairy and brutal. In this story, the mammoth appears as a "positive" figure, kindly protecting the heroes in a desperate fight against their enemies. The novel owes much to fairy tales, but also to myth: Naoh, the hero, is at the juncture of two powerful myths—that of Noah (he is a "just man" who saves humanity from total destruction) and that of Prometheus (he brings fire to mankind). In *Vamireh* the "abduction of Elem" seems to have been lifted from the *Iliad*.

These mythic tales are written in a colorful style that seems to reflect the vision of the "primitive people" involved. They combine the affectation of the Goncourt brothers' "artistic writing" with the exoticism of scientific names in conveying the sense that in those "savage ages," nature was strange and luxurious. Rosny Aîné's great talent consisted in creating an impression of fantastical savagery while using the concepts and even the very vocabulary of science. In his books, stylistic affectation joins scientific vocabulary, and scientific thought takes on a kind of lyrical density. Science becomes poetry, its language spoken in the measured cadences and complicated articulations of Latin nomenclature. In his prose, a primitive, barbarous chant is composed of words like "megaceros," "rhesus," "crustacean," "mastodon," and "proboscidean."

These novels condense dreams and scientific knowledge. As fiction attempts to apply scientific knowledge, does it "progress" along with it? The American novelist Jean Auel did a lot of research about the prehistoric people of central Europe and Russia before writing *The Mammoth Hunters* in 1985. Her story is set at the time of the warming period that occurred in that region between 35,000 and 22,000 B.P. (Wurm IV), and the description of the dwelling of these sedentary people of the Gravettian period is directly inspired by the work of Czech and Soviet archaeologists: "The side walls . . . seemed to be made of a mosaic of mammoth bones . . ." "a heavy curtain of mammoth hide covered the entryway. . . ." Objects that have been excavated are worked into the scene, for example, a musical instrument like a xylophone made of mammoth scapulae, or a mammoth bone spit used to roast meat: "A massive haunch of meat was cooking over it, spitted on a long pole. Each end was resting in a groove cut in the knee joint of an upright leg bone of a mammoth calf, sunk into the ground. . . ."[21] Set in the context of an encounter between Neanderthals and Gravettian *Homo sapiens*, the characters' adventures are related with considerable emphasis on romantic matters, including the sexual initiation of women and marital problems. The novel unfolds in a matriarchal setting that seems to reflect the values and theories of contemporary American feminism.

While some novels tried to re-create the life of prehistoric hunters, others brought the mammoth forward into the present. Among his tales of the far north, Jack London included "A Relic of the Pliocene," the saga of a mammoth hunt by an Alaskan hunter.[22] And in a novel published in 1928, Max Bégouën (one of the three boys who discovered the Trois Frères painted

An unexpected meeting between a mammoth and a dinosaur in a Japanese comic book.

cave in the Ariège region of the Pyrenees) imagines the "resurrection" of a mammoth in Soviet Siberia, thanks to electricity:

> The galvanic currents were applied to the nerve centers, as biogenic rays bathed everything in an intense purple light. . . . Suddenly, the animal's legs quivered. The bent trunk seemed to lengthen. The tiny eye, once dull and lifeless, appeared to swell and come alive, gazing vaguely about. The body temperature rose quickly . . . then the chest appeared to rise.
>
> "Oh, look," said Nadia, pointing at the faint cloud of steam coming out of the trunk's twin openings in rhythmic puffs.
>
> Nadia seized Mougin's hands. "It's alive," she cried.
>
> At that very moment, the mammoth's eyes blinked and a powerful blast of air roared from its rising trunk. . . .[23]

This isn't far from today's high-tech dreams of reconstituting "prehistoric monsters" thanks to computers and genetic engineering.[24]

Novels reflect the progress of knowledge and technique, but also the play of dreams and the imaginative richness of prehistory's themes. They reflect the interests of their authors and the public for which they are written. Brought back in this way to our contemporary preoccupations, the fiction of novels becomes more real and more authentic than the "reality" described by science. In some ways, fiction even takes the place of science. In the nineteenth century, scientists hid behind pseudonyms when they wrote prehistoric tales. Today they dub themselves "advisers" to prehistoric novels, films, and comic strips, and even write their own.

Movies often show "prehistoric monsters"—usually elephants in disguise—and life-size animated mammoths roam theme parks. But the comic strip remains paleontological fiction's highest form of expression. Written mainly for children, "prehistoric comic art" grew tremendously during the 1970s. Whatever its educational or moral intentions,[25] it usually traffics in images that are far from contemporary scientific knowledge and in which nineteenth-century clichés survive with amazing permanence. Some are based on even earlier models and ancient images of the "savage." The comics put

Tounga: Master of Mammoths (Brussels: Ed. du Lombard, 1982).

their heroes in a timeless setting unrelated to their appearance, environment, or technical skills. It is as if the immensity of geological time had been compressed and flattened into an abstract, shortened time span, making contemporaries of man and dinosaurs, ape-men and *Homo sapiens*, woolly mammoths and African savannah.

As a storytelling form, the comic strip seems especially well suited to presenting the fragmentary knowledge produced by the sciences of prehistory: linked by a narrative thread, its discontinuous images seem to embody the very way the shards of that knowledge were uncovered. But comics usually involve a story told at breakneck speed and crowded with a blur of events, where people run, fight, and hunt from the first page to the last. Nature's spasmodic violence is matched by the violence of the action, and the graphics—in a kind of expressionistic aesthetic—serve up a diet of exaggerated gestures, bent or twisted bodies, raised arms ready to strike, eyes wide in close-up, and gaping, drooling mouths with enormous teeth. In this world of nonstop disasters, with its monstrous or gigantic beings, cartoonists have somehow rediscovered the "formula" that made Cuvier's catastrophism such a success. In comic-book prehistory, time does not exist. The only time span that matters is that of the story and the book, not the immense sweep of geological time. Some of these tales, such as *Tounga*,[26] use traditional adventure-story narrative devices told against a threadbare "prehistory" setting. Others, like *Rahan*,[27] reflect such moral values as the dignity of the human species, brotherhood, mutual aid, and tolerance.

Here prehistory is but a font of images, as ape-men are thrown together with intelligent human beings in a luxuriant and violent natural world, where volcanic eruptions, tidal waves, and meteor showers are daily occurrences. In this vague, flattened past—which permits every narrative shortcut imaginable—australopithecines are the contemporaries of *Homo sapiens*, the invention of fire coexists with the sophisticated use of language, woolly mammoths walk alongside dinosaurs, and prehistoric life has been reduced to quasi-mythical themes.

To the contemporary imagination, the mammoth appears as an ambivalent figure. Its strangeness, huge size, and enormous tusks make it terrifying. But it is also touching, because it seems clumsy, as if trapped in its massive body. We know it used its threatening tusks to scrape the ground and break ice, and its big hairy trunk only to pick the buttercups it fed on. Mammoths that fell into traps or crevasses have been found dead, frozen in the ice, and we can't help but pity such awkward giants of the snows.

In a world where animals hardly exist any more for city dwellers except on television, a prehistoric animal is a figure of fiction, hovering between the nightmarish image of a monster and that of the gentle and inoffensive animal

A LITTLE CHRISTMAS DREAM.

Louis Figuier suggested educating children by telling them stories from prehistory instead of fairy tales, but is that such a good idea? Artist Georges du Maurier felt that prehistoric monsters would be more likely to give them nightmares. Ironically titled "A Little Christmas Dream," his cartoon appeared in Punch in 1868. (This illustration also appears in Martin Rudwick, *Scenes from Deep Time,* 1992, p. 43.)

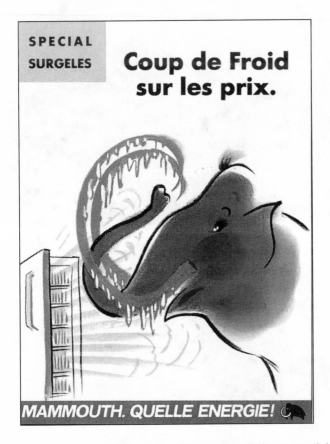

Images and symbols condensed: the mammoth in an advertisement for frozen food. (Publicity campaign for Mammouth stores, France, 1993.)

in cartoons that little children watch at bedtime. The mammoth has been adopted as the name and symbol of a chain of French supermarkets, whose advertising campaigns successively came up with various "concepts" related to the animal. They turned the mammoth into a symbol of power and size (it "tramples prices") or made it a more familiar and homey image that is protective, nourishing ("Mammoth, where you live"); picked up on the elephant's legendary intelligence (mammoth disguised as Sherlock Holmes); and—in an amazing compression of meaning and symbol—made the mammoth the advertising "spokes-mammal" for a line of frozen food.

Thus extinct animals reconstituted by the sciences of prehistory are being integrated in our world in various ways. Science becomes a reservoir of images by turn familiar, threatening, or strange, figures that are terrifying or touching, and that feed, for better or for worse, the stories, images, and dreams of our daily existence.

In France at the end of the twentieth century, a minister of education took the mammoth as a symbol of the heaviness of the French educational system. Demonstrations of students and professors against his reform aiming to "put the mammoth on a diet" forced the minister to resign in March 2000. Cartoon by Plantu published in *Le Monde* (March 30, 2000). Reproduced with permission of the author.

PART TWO

MYTHS

Saint Augustine
and the Giants

In book 15 of *The City of God*, Saint Augustine described an extraordinary discovery. A gigantic human molar, as big as a hundred modern teeth combined, was found on the shore at Utica, a city near Carthage on the Mediterranean coast of North Africa:

> I myself, along with some others, saw on the shore at Utica a
> man's molar tooth of such a size, that if it were cut down into
> teeth such as we have, a hundred, I fancy, could have been
> made out of it. I believe it belonged to some giant. For though
> the bodies of ordinary men were then larger than ours, the
> giants surpassed all in stature. And neither in our own age nor
> any other has there been altogether wanting instances of gigan-
> tic stature, though they may be few. The younger Pliny, a most
> learned man, maintains that the older the world becomes, the
> smaller will be the bodies of men. And he mentions that
> Homer in his poems often lamented the same decline; and this
> he does not laugh at as a poetical figment, but in his character
> of a recorder of natural wonders accepts it as historically true.
> But, as I said, the bones which are from time to time discov-
> ered prove the size of the bodies of the ancients, and will do so
> to future ages.[1]

What is this huge molar doing in the work of a fifth-century Christian theologian? And why did Saint Augustine (354–430) add that shortly after its discovery, the tooth was brought to a church so it could be shown to the faithful? For a father of the church to be interested in curiosities picked up

"A giant's tooth" (actually the fossilized lower molar of an elephant). Muséum d'Histoire Naturelle de Bordeaux. (Flammarion photo.)

on the beach, so be it. But that he mentioned it in an apologetic work titled *The City of God* seems a little incongruous.

In the chapter where Augustine wrote about the huge tooth, he explained that in the early days of the world, before the Flood, there lived a human race of gigantic size and of extraordinary longevity. With the passing of the centuries, men became smaller and their lives shorter, as if the world, in aging, lost its vigor. And since human size has diminished during the ages of the world, the huge bones and teeth that are sometimes dug up are the remains of this humanity from the earliest times of Creation and prove the dwindling of nature's primitive energy.

The belief in giants—which appears all over the world in myths, legends, folklore, and stories—takes on a particular dimension under Saint Augustine's pen. This traditional theme, this quasi-universal myth, here connects the representation of an origin and a history based on the biblical account.[2] All the events of history, past or present, belong to the sacred story and are aimed toward a single end. Such is the lesson of *The City of God*, a theological text and biblical commentary that appeared at a particularly dramatic moment in history.

Augustine, the bishop of Hippo, undertook to write the book in the year 413 as a defense of religion against the attacks of pagans following a great tragedy that struck the Christian world. Led by their chief, Alaric, the Visigoths invaded Rome on August 23, 410. The Holy City was pillaged and destroyed. Faith, prayer, and the presence of the apostles' relics were not enough to protect the faithful, and a number of Christians were forced to seek refuge in the cities of the North African coast. This grave event raised the question of the justification of evil and destruction in the face of Christian faith, and *The City of God* was written as a response to the distress that overwhelmed the faithful. To bolster religious hope, Augustine sketched a vast historical panorama and described the fates of two "cities" founded on opposite attitudes: "the earthly by the love of self, even to the contempt of God; the heavenly by the love of God, even to the contempt of self." Every act of human life is related to one or the other of these two inclinations.

> The glorious City of God is my theme. . . . I have undertaken
> its defense against those who prefer their own gods to the
> Founder of this city—a city surpassingly glorious, whether we
> view it as it still lives by faith in this fleeting course of time,
> and sojourns as a stranger in the midst of the ungodly, or as it
> shall dwell in the fixed stability of its eternal seat, which it now
> with patience waits for, expecting until righteousness shall re-
> turn unto judgment, and it obtain, by virtue of its excellence,
> final victory and perfect peace.[3]

The City of God is both a theological text and a manifesto of hope that promises the salvation of the just, the reign of good, and the triumph over evil "in the future age." Human history, said Augustine, unfolds in six successive ages, which, like the six days of Creation, will lead to a seventh epoch "marked by eternal rest, not only of the spirit but also of the body." Mankind is now in its sixth age and will soon enter the last one. Since the beginnings of humanity, one can see the interplay of this contradictory double aspiration, in Adam and Eve, in the conflict between Cain and Abel, and in the episode of the Flood. The Bible is therefore read as Holy Writ but also as a historical account from which Augustine drew arguments and evidence to support his demonstration. To his eyes, it could not be other than a *true* story: one can use it to establish precise chronologies of the world's history by studying the patriarchs' ages and genealogies, and establish the truth of every detail of the account.

Augustine's digression about giants comes into play to justify the fact that Cain was able to found "a city, and a large city." How could a single man found such a city? Only if men before the Flood lived far longer and were much bigger and stronger than men today does this biblical claim become plausible. Archaeological remains become evidence for the truth of Scripture.

"The large size of the primitive human body is often proved to the incredulous by the exposure of sepulchers, either through the wear of time or the violence of torrents or some accident, and in which bones of incredible size have been found or have rolled out," wrote Augustine. Perhaps the giant tooth had been sent by heaven to convince the "incredulous." Remains that are sometimes found on beaches or buried in the earth are the material link that connects us, we people of the last age, with those of the youth of the world.[4]

It is therefore understandable that the molar found by Saint Augustine on the Utica beach should have been displayed in a church: it was a relic of the most ancient humanity, close to the earliest times of Creation. Fossil objects were divine signs. For Saint Augustine, nature as a whole was miraculous because it was seen as the expression of divine will. All the objects of the world were prodigies; only our familiarity with them has dulled our sense of wonder.[5]

The "giants" doctrine also found support in the writings of poets and naturalists of pagan antiquity. Didn't the "learned Pliny" assure us that "the longer time flows along its path, the more bodies produced by nature diminish"?[6] During the Middle Ages and the Renaissance, Pliny's *Natural History* was a major reference work.[7] Just as Pliny quoted Homer on this subject, not "as a poetical figment, but in his character of a recorder of natural wonders accepts it as historically true," Augustine also used Virgil, Pliny, and Homer to confirm the decline in gigantism of peoples of the past:

> When the most esteemed of their own poets, Homer . . .
> speaks of that huge stone which had been fixed as landmark,
> and which a strong man of those ancient times snatched up as
> he fought, and ran, and hurled, and cast it—and Virgil adds,
> "Scarce twelve men of later mould / That weight could on
> their necks uphold," thus declaring his opinion that the earth
> then produced mightier men.[8]

And just as Virgil brought up the ancient memory of those Homeric heroes and their enormous stature, so all the more reason that even older times should have witnessed more gigantic beings "in the ages before the world-renowned Deluge"?[9]

The huge tooth found on the Utica shore was probably that of a fossil elephant. Since Greek and Roman times, these pachyderms' fossilized skulls and teeth, which are often found on Mediterranean shores, had been interpreted as proving the existence of mythological giants.[10] From antiquity, it was widely believed that the famous Cyclops were the first inhabitants of Sicily, as evidenced by the many gigantic bones found in the soil and buried in caves there. These were probably the skeletons of now-extinct dwarf elephants, whose remains are found in abundance in the island's Quaternary layers. The nasal cavity of these pachyderms was thought to be the Cyclops' "single orbit."[11]

During the Middle Ages and the Renaissance, many works reaffirmed the belief in giants. These "gigantologies" were based on the Bible and church fathers like Saint Augustine or Isidore of Seville, on Greek and Latin authors of antiquity, and on many ancient or recent discoveries of the remains of "giants."[12] These include a gigantic skeleton found at Tegea, said to be that of Orestes and that is mentioned in Herodotus (I:67–68); another found in Crete after an earthquake, according to Pliny;[13] and the skeleton of Antaeus was discovered in Tangier, according to chapter 13 of Plutarch's *Lives*. In Boccaccio's *Genealogies of the Pagan Gods*, he described a 1342 discovery made in Sicily in the Erice Cave near Trapani. Digging the foundations for a house exposed a cavern in which were buried the bones of a man "two hundred cubits" (about four hundred feet) tall, seated in the cavern and holding a club "longer and thicker than the mast of a ship." The skeleton crumbled into dust as soon as it was touched, however; only some bone fragments and three enormous teeth remained. "To display their discovery to posterity, the citizens of Trapani strung [the bones] on a wire and carried them to a church in the city dedicated to the honor of the Annunciation."[14] Boccaccio thought the remains might be those of the giant Polyphemus.

This last story is not just another fable from the facetious author of the *Decameron;* it raises specific issues related to fossil objects and their interpretation. It directly connects ancient history (or mythology) to concrete evidence, that is, excavated bones. And it relates the theme of giants to religious beliefs and institutions. These bones of "the giant Polyphemus" were exhibited in a church—which in the mid-fourteenth century was still "the place to preserve and display signs from the past and miraculous and/or marvelous objects."[15] Like the relics of saints, gigantic bones and teeth found their way to the churches and abbeys' "chamber of marvels." "In the Middle Ages, everything that evoked wonder and which seemed rare and unusual was brought together inside the church": relics and fossils, but also precious stones, antiquities, and exotic objects. Thanks to these treasures, "which were highly effective propaganda devices, a church or ecclesiastical community acquired a reputation, and, as a result, could accumulate great wealth through pilgrims' offerings."[16] The teeth or bones of "giants" were both curiosities and evidence of the miraculous. Until the seventeenth century, it was not unusual to see them publicly displayed as being the remains of Saint Christopher or some other patron saint, and borne in procession on feast days. As late as 1789, "in praying for rain, the canons of Saint Vincent's would parade through the streets and countryside carrying what was believed to be a saint's arm, but in fact was the femur of an elephant."[17]

Between the fourteenth and sixteenth centuries, giants flourished in folklore and literature. Rabelais' burlesque giants Gargantua and Pantagruel mocked traditional beliefs and biblical genealogies. Knowledge about giants became the subject of debates over both the theological dimension of history and the question of the nature of fossil objects. Were sacred texts to be taken literally or interpreted metaphorically? "The theme of giants is at the heart of questions raised by the thesis of the aging of the world," writes Jean Céard.[18] Giant lore meshed with naturalists' preoccupation and interest in archaeology and curiosities. Giants' teeth and bones had pride of place in collectors' cabinets of curiosities. In sixteenth-century Italy, they were found in the collections of Philippe Costa, a gentleman from Mantua; the Venetian Giorgio Cavazza; and, in Rome, Prince Federico Cesi, founder of the Academia Nazionale dei Lincei, to which Galileo belonged. In France in 1642, Pierre Borel of Castres boasted that his cabinet contained "a giant's scapula, or shoulder blade, that weighed 35 pounds, was four paces high and seven wide," as well as "two giant teeth, half the size of a fist."[19]

At the same time, people were speculating on the nature of fossil objects. During the Renaissance and until the seventeenth century, the term "fossil" referred generally to all objects found in the earth, with no distinction made between native minerals and organic remains. By tradition, medieval lapidaries and Renaissance encyclopedists classified under a single heading gems, minerals, remains of living creatures, shells, bones or horns, and other objects, whether fantastical or handcrafted, found buried in the earth. "It should be known that fossil means everything that can be drawn from the bosom of the earth by digging," wrote a naturalist at the end of the Renaissance.[20]

But what exactly were those "giants'" teeth and skeletons, and the fossil shells and fishes found in the earth and sometimes even at the tops of mountains? Were they "essential" (that is, formed in the earth) or "accidental" fossils? Were they the images of things, "sports of nature" imitating the shape of living creatures, or the remains of those creatures themselves? Were they sui generis products of the earth by virtue of a "lapidific fluid," or should they be taken to be actual living beings that had been "petrified"? In Neoplatonic thought, the subterranean world can create images of beings modeled on those of the aerial world, in a kind of correspondence linking the microcosm of the human body to the macrocosm of the world.[21] These ideas, borrowed from the naturalists of antiquity, were picked up by such Renaissance encyclopedists as Conrad Gesner in Germany and Ulisse Aldrovandi in Italy. In Aldrovandi's nomenclature, "fossil objects" were given names according to what they looked like, with the Greek root referring to the living thing being imitated and the suffix "-ite" referring to those objects' "stony" substance. In his *Museum metallicum*,[22] each stone had a name, attributes, and distinctive properties according to what it resembled. Thus, Aldrovandi distinguished between *ophiomorphites* (stones shaped like snakes), *sepites* (like squid bones), *psetites* (fish), *echinites* (sea urchins), "cancriform stones" (crabs), and *ostracites* (oysters). He also had stones bearing images and inscriptions; one showed the picture in stone "of a group of Tartars with their sheep and camels near Samogedi." In his *Mundus subterraneus*, published in 1665, the Jesuit Athanasius Kircher claimed to have found carved or sculpted in fossil stones a complete Latin alphabet as well as images of the Virgin and Child.[23]

In France and Italy, however, those beliefs had long been the object of serious criticism. At the end of the fifteenth century, Leonardo da Vinci refused to believe that "figured stones" were born on hilltops under the influence of stars. "Where in the hills are the stars now forming shells of distinct ages and species?" he asked.[24] "Fossil shells" were at the very heart of the debates. The presence of these marine vestiges in the earth and, sometimes, on mountaintops, was a real enigma, if, like most Westerners of the time, one believed that the world had no history and appeared to our eyes as it had been created by God in the very beginning.

In the first years of the sixteenth century, the Italian naturalist Girolamo Fracastoro studied fossil shells that had been dug up during excavations in Verona in 1517 and claimed that they belonged to animals that had lived there. In France at the end of the sixteenth century, a Saintonge potter named Bernard Palissy identified fossils as animals that had once existed. "Looking carefully at the shapes of the stones, I found that none could have taken the shape of a shell nor that of any other animal, had the animal itself not created the shape."[25]

Palissy had gathered fossils in Saintonge and the Ardennes, and in the lectures he gave in Paris in 1575, he rejected the idea that they were "sports of nature" or "figured stones." He claimed that the fossil layers had been formed at the bottom of the seas: "[I] maintain that crustaceans that are [found] petrified in many quarries were created at the very place, when the rocks were but water and mud, and which have become petrified along with the selfsame fishes."[26]

The "bones of giants" were also interpreted as "figured stones," "sports of nature," or the remains of elephants and whales. By 1560 the giants hypothesis was but one among many. In 1550 T. Fazello was still busy defending it against skeptics,[27] while in Holland Goropius maintained in his *Gigantomachie* that the so-called giants of the Bible and antiquity were no taller than people today.[28] In these debates, one can see the crisis of the theological representation of the world's history and the emergence of a new way of looking at nature and life. All these elements would come into play between 1613 and 1618 in the famous affair of the "giant Teutobochus."

In 1613 some bones of extraordinary size were found in a sand quarry near the castle of the counts of Langon in Dauphiné, in southeastern France. The discovery was announced by Pierre Mazurier, a master surgeon at Beaurepaire, who claimed they were the bones of a giant thirty feet tall. There were ten in all, "to wit, two pieces of the lower jaw, two vertebrae, part of a rib, the neck of a scapula sinister, the head of an arm, the head of the thigh, the leg, the astragalus, and the heel, all from the left side."[29]

The bones were displayed in many cities of France and Germany, exhibited at fairs, in marketplaces and town hall squares. The giant's skull and bones must have been a sight to behold. The remains were even presented to King Louis XIII, who marveled at them. To spread word of his discovery, Mazurier asked a certain Jean Tissot, a Jesuit in Tournon, to write a brochure, which was published in Lyon in 1613. It was learnedly titled *A truthful account of the life, death, and bones of the giant Teutobochus, king of the Teutons, the Cimbri, and the Ambroni who died 105 years . . . before Jesus Christ . . . and was buried near a castle named Chaumont now called Langon, in Dauphiné, where his tomb was discovered.*

Mazurier was sure the bones were those of the giant Teutobochus because, in addition to the bones, a sepulcher had been found "made of bricks, well cemented in its four parts, being 30 feet long, 12 feet wide, and 8 feet deep, including the headstone, whose center bore a gray stone into which the epitaph 'Teutobochus Rex' had been carved."[30]

Teutobochus is well known to the historians of antiquity, and Jules Michelet told the legend in his *History of France.* Teutobochus was the king of those barbarous northern peoples—the Teutons, the Cimbri, and the Ambroni—who, "with their enormous stature, fierce gaze, and strange weapons and clothing," had fought the terrified Romans right to their city walls. "Their king Teutobochus was able to jump over four or even six horses. When he was brought in triumph to Rome, he stood taller than the trophies."[31] Paulus Orosius's *Historiarum adversus paganos,* of which a new translation from the Latin had been published in French in 1509,[32] said that "one thousand eight hundred years ago" this northern giant had led an army of seven hundred thousand men to Dauphiné, hoping to "trample the Roman Empire." A battle took place between the armies of the Teutons and the Cimbri and the Roman legions commanded by Marius, in which the Romans were victorious. Teutobochus was killed in 101 B.C. and buried in a place whose name has been confirmed by tradition from that distant past: "Ever since the great battle, people of Dauphiné, from father to son, have called the place where the sepulcher was found, the Giant's Field."[33] Finally, the epitaph "written in Roman letters in a stone," "Teutobochus rex," and "several silver medallions . . . bearing Marius's image on one side and a large entwined M and A on the other," which were found next to the bones when they were discovered, served as reminders of Marius's victory over the barbarian army.

The discovery of this "tomb" provoked contradictory reactions. Some denied that the bones could really be bones, choosing to see them as "sports of nature" or "figured stones." Others viewed the Teutobochus thesis with suspicion, doubting that the bones were that of a man, and preferring to believe that they were the remains of some gigantic animal. The discovery raised at once problems of anatomy, natural history, and theology. The bones were sent to Paris to be examined, where Nicholas Habicot, "a master surgeon at the University of Paris,"[34] was assigned to study them. After examining the bones, Habicot published in 1613 a sixty-three-page pamphlet called *Gigantostéologie, ou Discours des os d'un géant* (Gigantostology, or An account of a giant's bones) with Jean Houzé at the Palais in the Galerie des Prisonniers. In his pamphlet, Habicot confirmed Mazurier's thesis and expanded on it with his own anatomical and theological arguments. This marked the start of a six-year war of pamphlets, tracts, and at least a dozen treatises. That same year, a document called *Gigantomachy, a Response to Gigantostéologie* was published anonymously, signed only "a student of medicine." Its author was Jean Riolan, a young botanist who would become the most brilliant anatomist of his time and the personal physician to Marie de Médicis.[35]

This quarrel was no mere academic dispute, opposing the traditional knowledge of "surgeons" with that of the developing sciences of anatomy and medicine. It crystallized all the polemics about fossils that had taken place throughout the Renaissance. In the years when the scientific revolution was taking shape, it could be seen as the battlefield where supporters and adversaries of giants faced each other in a final combat, the moment when the view of fossils would free itself from the myth of giants—not from a universal myth, but from a particular aspect of myth born of traditional exegesis of sacred texts.

So one should not focus only on the naive or picturesque parts of this "gigantomachy." The debate should be seen in the context of the time, when the question of the nature of "fossils" was being furiously debated, and a rational approach to the understanding of nature becoming mandatory. The existence of buried proboscidean remains challenged the knowledge of an entire era: theology, philosophy, natural history, and alchemy. It also brought out the various narratives used to explain fossils, and, through their interpretation, the representation of nature and history at the dawn of the modern age.

Habicot's anatomical knowledge derived mainly from his practical skill in osteology as a military surgeon. His first task was to identify the Dauphiné fossils as being actual bones. For Habicot, the fossils were certainly not sports of nature. If, he wrote, "the teeth and the jaw are as if petrified," that did not mean that they were not originally bones. It is possible to imitate the shape of bones, but not their substance, and in the very substance of our giant's bones, one can distinguish "porosities, fibers, and cavities, which no artist could duplicate."[36]

Some of the bones, once brought to light, had crumbled to dust. Others had been naturally petrified by water flowing in that location. Doesn't the marvelous preservation of saints' bones prove how well bones can be preserved? "I have been shown a calcaneum [heel bone] from Saint Peter's foot, and Saint Laurent's big toe bone, which was somewhat grayish, due to the fire," wrote Habicot in all seriousness, in defending his thesis.[37] Some people believed that the bones, "of such frightening size," may have belonged to animals like the elephant or the whale, whose skeleton is enormous, "as one can judge from the ribs one sees on the right when entering the church of Saint-Denis in France." But Habicot firmly rejected this argument. "The elephant and the whale have bones made in another manner" than those of man. Besides, he wrote, "man has a soul," and his nature is profoundly different from that of animals, which makes any comparison a priori impossible. To the contrary, he identified the bones of his "giant" with specific reference to the human skeleton, examining the teeth, vertebrae, "the neck of the scapula," "the head of the left femur," the tibia, and the calcaneum. Habicot had no doubt as to his conclusion: the Dauphiné bones were indeed those of a human being of exceptional size.

In making his case, Habicot didn't hesitate to draw on theological arguments, taken directly from "Holy Writ, the books of Genesis, Numbers, Deuteronomy, Joshua, Judges, Samuel, the Chronicles, and Baruch." After all, doesn't the Bible tell us that Samson wrestled a lion to death and tore off the gates of Gaza?[38] One must conclude, with Saint Augustine and the church fathers, that men were taller before the Flood and that humanity has been shrinking. Giants became rarer because "physical beauty and tall stature" were less desirable than "knowledge of the way of science." That is why one doesn't see giants very often nowadays, and why they are, well, *smaller* than those of the past.

To buttress his demonstration, Habicot invoked scientists and poets of antiquity, ancient mythology, legends, and the history of France. "France's chronicles and annals of the life of King Charlemagne relate that his nephew Roland killed a giant named Ferragut, who was twelve cubits, or thirty feet, tall. His face was a foot and half long; his nose, ten inches; his thighs, four cubits." Habicot also used hearsay about curiosities he or others had seen. "I myself saw, at the late Madame de Nemours's, a man more than fifteen feet tall, and several people in this century have seen in this city a Flemish man who was nearly as tall." Finally, the remains of the Dauphiné "giant" were those of an individual with a particular historical identity, and must therefore be related to history, and in particular to Roman history.

To Tissot's claims, Habicot therefore added his own anatomical knowledge, bolstered by some scattershot erudition. What strikes us as naive and burlesque about his demonstration is the coexistence of so many arguments of different kinds, all located at the same level of truth: anatomical evidence, erudition about antiquity, mythology and history, the witnessing of curiosities, hearsay, and historical and archaeological evidence, not to mention theological arguments, epigrams, sonnets, octets, and Latin verses penned by various witnesses. In the early seventeenth century, this process of accumulation represented an approach to knowledge still typical of the medieval scholastic tradition, which joined scriptural truth with scholarly erudition from the ancients, and, in medicine, from those "pedants" whom Cyrano and Molière would lampoon a few decades later.

Riolan criticized Habicot's efforts harshly, calling the identification of the giant's remains and King Teutobochus's tomb a fraud, or at the very least a mistake. After a reply from Habicot (which essentially traded insult for insult),[39] Riolan in 1618 published a new pamphlet[40] that seemed to place him in the same register as his adversary: the same bookish culture, the same erudition, the same references to the Bible and to the authors of antiquity, to history and to archaeology. But Riolan's approach was different, because truth for him was to be found above all in observation. Invoking Galen and Hippocrates, he affirmed that many "falsehoods and absurdities" had been "handed down in medicine" without sufficient evidence. "Those who published them first, out of negligence or excessive credulity, did not bother to recognize the truth. Those who came later relied on the accounts of the first. In this way, from hand to hand and from book to book, falsehoods and absurdities have been given authority and received as truthful."[41]

Unlike Habicot, Riolan thought tradition could be misleading. What was now involved was the actual means of verifying and accumulating knowledge—knowledge that Habicot received and transmitted uncritically, but that Riolan confirmed and verified by observation and experience. In some ways, these two protagonists embody two opposite attitudes toward science, its methods and constitution.

One by one, Riolan refuted Habicot's religious, historical, and anatomical arguments. He smashed the theological framework of his adversary's speculations, starting with the idea that the world is aging—an idea he cautiously attributed not to the Bible, but to the Epicureans.

> I know that the Epicureans claim that the world continues to
> age, and that the energy of elements is being slowly consumed,
> with the result that animals are getting smaller. But this reason-

ing by the Epicureans is frivolous. If the world were aging, it would no longer produce offspring, since age is recognized by sterility and the inability to reproduce, and not by the small size of animals. And if it were true of animals, why should it not also be noticed in plants?[42]

"The tallest men of the first and second ages were no more than nine or ten feet tall," he added.[43] Where Habicot and Tissot related the anatomical explanation to mythology, history, and theology, Riolan mainly seemed concerned with identifying the remains by anatomical examination; his approach was closer to what we would consider today that of a naturalist. Still, the hypotheses he advanced have a certain strangeness in our eyes, for example,

GIGANTOLOGIE

HISTOIRE
DE LA GRANDEVR
DES GEANTS,

Où il est demonstré, que de toute
ancienneté les plus grands hommes, & Geants, n'ont esté plus
hauts que ceux de ce
temps.

*Quis autem vestrum assiduè cogitans po-
test adijcere ad staturam suam cubi-
tum vnum?*Matthei cap.6.

A PARIS,
Chez ADRIAN PERIER, ruë Sainct
Iacques,
M. DC. XVIII.

One of the many (anonymous) pamphlets in the "Giants quarrel"
(1618). Its author is Jean Riolan, "a student of medicine." (© MNHN
Paléontologie, D. Serrette photo.)

when he considered—invoking Pliny—that some strange marine creature may be involved, "of a perfect or imperfect human shape, such as tritons, nereids, or sirens," "of enormous size."[44] Early in the seventeenth century, the notion that the sea could be hiding fabulous monsters in its bosom was a plausible hypothesis and would remain so during much of the eighteenth.

The Dauphiné remains could be those of such "marine monsters," but they could also be those of whales, an argument supported by many examples. Bones of these marine animals were often displayed in churches and monasteries. "The great bones in the holy chapel of Bourges . . . were brought from Dauphiné"; and "in Valence, a Dauphiné city, large bones can been seen at the Cordeliers monastery."[45] People have also seen "a tooth a foot long" that "weighed eight pounds," and "no one knew what animal could have produced [those remains]." In short, the countryside "in the mountain valleys close to the Rhône River is full of such bones." Travelers' tales brought other accounts. Wrote Pliny: "The commanders of the fleets of Alexander the Great have related that the Gedrosi, who dwell on the banks of the river Arabis, are in the habit of making the doors of their houses with the jawbones of fishes, and raftering the roofs with their bones." The whale hypothesis would also explain the molar found by Saint Augustine, the bones of the so-called "Boccaccio giant," those of the giant Pallas, and the Homeric legend of the Cyclops.

In his next chapter, however, Riolan proposed to identify the bones of the giant as those of elephants—with apparently equal conviction. "I freely confess that I have never seen an elephant," he wrote, "much less observed or studied its bones, so as to know in what ways they are similar to and dissimilar from human bones." In the early seventeenth century, elephants were not well-known in Europe, and Riolan's knowledge could only have come from books. Yet his demonstration illuminated the overall resemblance between elephant and human skeletons and their differences in detail: "The elephant, as has been said, has heels on its hind legs; its posterior legs are longer than the anterior. . . . It bends its rear legs the way man does, unlike all other animals. . . . Its foot is rounded, like a large basin . . . and the bones inside divided into five toes."[46]

An elephant leg, like a human one, ends in five toes, which is one of a number of analogies and differences Riolan noted. Like man, the elephant "has two large condyles on the lower part, and between the two is a cavity, where the kneecap is located. One can also see the two trochanters, but they are of different shape and location; the neck of the thigh is quite short, and not curved." The bone structures of elephant and human members are similar enough, Riolan suggested, that they could be confused. "It is conceivable that the remains of elephant bones, except for the head, resemble those of man, and that it would be very difficult for a doctor or surgeon to tell them apart, unless he were a good anatomist."[47]

What was particularly novel about Riolan's approach is that a seventeenth-century anatomist should be making such a comparison while studying fossil bones. One can see why his work earned the admiration of Georges Cuvier, the French founder of comparative anatomy, two hundred years later.[48]

And yet, one large enigma remained: Where did all those elephants come from, whose remains were being found pretty much everywhere in the layers of the earth? The answer was simple. They were brought there by Hannibal's army. Elephants were robbing the giants of their mystery. Once again, Roman history was trotted out to explain the incongruous presence of these buried bones. The belief that the remains of elephants found in the upper layers of European soil—especially in Italy and in France in the Rhône Valley—were those of animals brought from Africa by the Roman armies became widespread in the sixteenth century and remained a sufficiently popular notion until the late eighteenth, that in 1812 Cuvier deemed it worth refuting.

"A careful study of the authors who have described Hannibal's march should have noted this error, even before one knew the circumstances under which the bones were found," Cuvier wrote.

> In fact, Hannibal only brought 37 elephants to Italy (*Europ. brev.* III, chap. vii), and Polybius tells us that all but one died of cold immediately after the battle of Trebbias. Livy, who is more accurate, says Hannibal had 8 elephants left, 7 of whom soon died during his unsuccessful attempt to cross the Apennines in winter. But the two other writers agree that in the spring, when Hannibal descended into the swamps of the lower Arno, he only had a single elephant left. The general rode it for this difficult crossing, during which he lost an eye to an inflammation. It is perfectly obvious, as Targioni and Nesti have pointed out, that this single elephant could not have produced the vast quantities of bones that are scattered all over Tuscany. Moreover, we know today that there are almost as many hippopotamuses as elephants, and that the two are often found jumbled together in the same layers. It is simply not credible that the bones could have come from animals used in warfare.[49]

In the third part of his demonstration, titled "Of Fossil Bones," Riolan examined "the fact that bonelike stones, which resemble human bones, can be created and formed in the earth."[50] He then reviewed the various doctrines used to explain the generation of fossils in the earth. The supposed "bones of giants" could be but figured stones. Hadn't Gesner "told of so many resemblances between stones and animals and artificial things, that no one could doubt that bones similar to ours could be created and formed in the earth?"[51] If "bones and stones" can be created within our bodies, why couldn't the earth, "our common mother," give birth to "stones similar to human bones?" For that matter, we have seen formed in the earth "the shameful parts of women," "clasped hands," stones resembling the "human thumb" or brain . . . and even teeth "so excessively large that it is inconceivable that there could have been men or animals of such a size." These bones could well be "sports of nature," and Riolan brought up all the traditional explanations inherited from Pliny and alchemical knowledge to support this new hypothesis.

Without choosing a single answer, Riolan juxtaposed all the various hypotheses that could account for the presence of the excavated bones. "I say that these bones are the bones of marine monsters in human shape, or the bones of a whale or an elephant, or else fossil bones."[52] Should we take this accumulation of contradictory hypotheses as a "weakness"[53] or rather a way of constructing an explanation that is no longer familiar to us? Marine monsters, elephants, whales, fossils created by the earth, all coexist in Riolan's text to refute Habicot's and Tissot's theologico-archaeological theses. It is as if these "hypotheses"—in the pre-Galilean sense of the word[54]—were all equally plausible and made meaningful by their very profusion. But the interest of this document lies precisely in that it gives us a fairly complete picture of the explanatory framework within which fossil objects could be placed at the end of the Renaissance. The giants thesis had been seriously undermined over more than half a century. What remained were Aristotelian or alchemical theories of the "formative power" of the earth and the germination of stones, and that of "petrification" of the remains of beings that lived in the past. Riolan did not choose to identify the skeletons of these supposed giants as being those of whales or elephants, or as objects created by the earth, but he did deny the fossils their status as divine signs and rejected the restrictive biblical framing of the world's history. And yet the history that our two protagonists drew on remained within the chronological bounds of biblical or Roman history—that is, human history. It would be up to the next generation—beginning in midcentury with Descartes—to produce a new kind of narrative that would expand the earth's history to cosmic dimensions.

Which leaves us with the skeleton of Teutobochus the giant. Today, at the entrance to the paleontology gallery of the Paris museum, you can still see a glass case bearing the intriguing label "Bone from the giant Teutobochus preserved at the Château of Langon." Was this, as Cuvier thought, part of a mammoth skeleton? Or the remains of a mastodon, as Blainville claimed in 1832? Or was it that of a *Dinotherium*, a primitive proboscidean whose tusks curved downward from its lower jaw, as has been more recently affirmed?[55] Whatever the case, these various hypotheses agree that this gigantic being was a member of the Proboscidea order.

The extraordinary bones of the supposed giant Teutobochus were displayed at fairs for a fee. The Dauphiné remains hadn't really been buried in a tomb, of course. Were they one of the many forgeries in the history of archaeology? Gassendi claimed that the "Roman" medallions in fact bore Gothic characters.[56] But archaeological and paleontological layers could have been accidentally mixed, producing the fortuitous association of remains of a fossil proboscidean with more recent archaeological vestiges. If it was a forgery, it was a profitable one, and it had the merit of provoking a debate that cast serious doubt on the belief that huge fossil bones were those of men from the very beginnings of the world.

During the first third of the seventeenth century, the enormous bones, huge teeth, and ivory "horns" being dug up were finally attributed to animals, and not to gigantic people. From the limited time span between Genesis and the Last Judgment in the Christian tradition, the shift was toward a sense of limitless time whose outcome is not determined in advance. "The world is not aging, it is eternally young, which allows it to create and re-create itself indefinitely."[57]

Giants did not abruptly vanish from the universe of knowledge, however. At the end of the century, the Jesuit Athanasius Kircher listed the traditional litany of famous giants from the Bible, from antiquity, and from more recent history.[58] He mentioned Goropius's Antwerp giant, Og and Goliath, the Cretan giant cited by Pliny, that of Fulgose, the famous "coffin of Orestes, which an oracle ordered reburied, and which measured seven cubits," and the Erice giant described by Boccaccio. But Kircher criticized as unlikely these traditional giants' excessive size, and carefully compared their heights.

	cubits	feet	hands
Height of Goliath	6½	13	19
The body of the giant found in Switzerland	9	18	27
The exhumed body of Orestes (per Pliny)	7	14	21
The body of Asterius, son of Anactis (per Pliny)	10	20	30
The body dug up in the mountains of Crete (per Pliny)	46	92	138
The body of the Tangiers giant, in Mauritania (per Pliny)	60	120	180
Lastly, the one that stretches credulity, found seated, in Erice, on Mount Drepano [Trapani] (per Boccaccio)	200	400	138

The Jesuit Athanasius Kircher protested the exaggerated size that his contemporaries ascribed to giants in *Mundus subterraneus* (1665), 2:56. (Bibliothèque Nationale de France photo, Paris.)

"The giants invented by the most recent authors are exaggerated," Kircher concluded. He was especially critical of the unbelievable size of Boccaccio's giant and suggested that its height should be reduced from four hundred feet to thirty. . . . At the end of the seventeenth century, giants were getting smaller. The horizon of what was believable was shrinking. At the same time, Leibniz refused to accept the existence of giants for "mechanical" reasons because, he said, "according to Galileo's reasoning, the size of animals must have limits."[59]

Gigantism was also becoming secular. Though the remains of giants continued[60] to fill the cabinets of the curious[61] and the writings of naturalists, fossils were losing their significance as divine signs. Giants were now only part of teratology, accounts of travel to such distant places as China or Patagonia,[62] folklore, or fairy tales. The theme of giants runs all through the eighteenth century and remains linked to fossils and the history of the earth. Buffon, in his *Epochs of Nature*, believed in the gigantism of the world's first inhabitants, though less out of religious faith than from a supposedly scientific explanation of the origin and history of the world. A few decades later, Cuvier would prove that these huge bones belonged to large extinct animals.

And yet, the myth of giants lives on in scientific paleontology. At the beginning of the twentieth century, the American naturalist Henry Fairfield Osborn thought that in the course of phylogenic history, species tended to grow in size and then shrink as an effect of aging.[63] The giants of nature, such as the *Titanotherae* or gigantic proboscideans, disappeared because they either became smaller or became sterile. Those two explanations, which were current in the seventeenth century, would endure in biology until early in the twentieth and even beyond. In 1940 the great paleoanthropologist Franz Weidenreich, a German refugee to the United States, smilingly referred to the biblical tradition in defending the idea that the origin of humankind might be found in a huge primate, the *Gigantopithecus*.[64]

Leibniz's Unicorn

A marvelous animal is said to exist that looks like a white horse, with a single, straight horn jutting from its forehead, and which can only be captured by a young virgin.

The medieval legend of the unicorn was probably born during Alexandrian times in the first centuries of the Christian era. The second book of the *Cyranides*, a text attributed to Hermes Trismigestus,[1] mentions this extremely amorous animal, which only a very beautiful woman can catch. In a strange reversal, this erotic myth may be at the origin of a long Christian and humanist tradition that turned the unicorn into a symbol of mystical purity.[2] Many tales have sprung from this legend, and medieval tapestries—such as *La Dame à la licorne* at the Musée de Cluny in Paris and *The Hunt of the Unicorn* at the Cloisters in New York—perpetuate its poetic image.

Where is this extraordinary animal? Did it ever exist? Does it still? Naturalists and travelers debated the matter throughout the Renaissance.[3] In those days unicorn horn was one of the most precious ornaments in cabinets of curiosities and royal treasuries.[4] Whole, in pieces, or ground up, it was worth more than its weight in gold because it was renowned for its marvelous medicinal properties:[5] it was a remedy for heart disease, an aphrodisiac, and especially, in powdered form, a highly effective antidote to poison. This is why unicorn horn was an essential ingredient at the table of princes during troubled periods of history. In the second half of the sixteenth century, no Italian nobleman would be without a fragment of this precious material. But the origins and properties of unicorn horn were hotly debated. In a veritable "battle of books" between 1550 and 1700,[6] no fewer than twenty-five papers appeared, quoting each other and repeating the same arguments and sources ad infinitum. Scholars, encyclopedists, travelers, doctors, and apothecaries expounded on the subject.[7] Ambroise Paré was categorical: "One must . . . believe," he assured, "that unicorns exist."[8]

But in the middle of the seventeenth century, Danish naturalists traveling in the far north identified the spirally twisted horn (as the unicorn's single horn was often portrayed) as the asymmetrical tooth of a narwhal.[9] So unicorn horn gained in banality as it lost its value. But "fossil unicorn [horn]" was also being dug up that was just as effective against poison. In the second half of the seventeenth century, the name *unicornu fossile* was given to a variety of whitish, oblong substances that were roughly horn shaped and found in the ground: the long bone fragments of great quadrupeds, pieces of wood, and fossil ivory tusks. Some "unicorn horns" on display in churches were curved. The church of Halle, Germany, today displays a mammoth tusk on a wrought-iron stand decorated with unicorns.

The Jesuit priest Athanasius Kircher wrote learnedly on the nature and origin of the famous "fossil unicorn" in his *Mundus subterraneus*—published in Latin in Rome in 1665—which was a late synthesis of the knowledge of the Middle Ages and the Renaissance.

Unicorns in a seventeenth-century engraving. (Edimedia photo.)

Nothing in nature has more value to the emperors, kings, princes, and great ones of this world than unicorn horn, to the point where even gold and precious stones are nothing in comparison. To what animal this horn belongs, no one can say. So far, neither doctors, naturalists, nor explorers know. Yet I say that if one denies that such an animal, whose horn has so many miraculous virtues, exists or has existed, one just as brazenly casts doubt on many pages of Holy Writ and on the truthfulness of all historians. It is certain that the unicorn, or *Monoceros*, exists.[10]

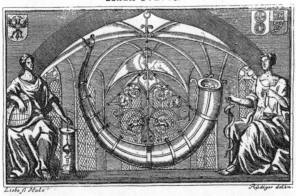

Title page of a work on fossil unicorn (1734). The mammoth tusk—presented here as a "fossil unicorn" horn—hangs in the church at Halle, Germany, in a wrought-iron frame decorated with unicorns.

The origin and nature of unicorn horn remained the subject of passionate arguments at the end of the seventeenth century. Not surprisingly, one finds the echo of those debates among those working to solve the enigma of fossil objects' nature and meaning, such as naturalists and philosophers, including the great Gottfried Wilhelm von Leibniz.

A curious reader leafing through Leibniz's small treatise *Protogaea, or A Dissertation on the Original Aspect of the Earth and the Vestiges of Its Very Ancient History in the Monuments of Nature*,[11] published in Latin in Göttingen in 1749, might be brought up short by a strange image. It appears at the end of the book, following several engravings showing fossilized fishes, shells, and fossil teeth, which were common in the kind of works of curiosities and natural history that became widespread at the start of the eighteenth century. The last page of illustrations has two engravings. One shows an enormous tooth, identified as that of a "marine animal." The other shows (in side view) the skeleton of an animal standing on two front legs, its spine slanting down toward the ground and ending in a kind of tail. The other end bears a horselike head topped by a single, straight horn. A strange skeleton of a two-legged unicorn! The engraving is captioned "Image of a skeleton excavated near Quedlinburg" (*Figura sceleti prope Quedlinburgum effossi*).[12]

Does this mean that Leibniz—the greatest German philosopher of his time, the rationalistic disciple of René Descartes, the brilliant inventor of differential calculus—believed in unicorns?

Protogaea, which Leibniz probably began in 1691, remained unpublished during its author's lifetime. But he did publish a summary of it in 1693 in *Acta Eruditorium*, the scientific journal he founded in Leipzig in 1682. More than fifty years later (and thirty-three years after its author's death), the original Latin text was edited by Ludwig Scheidt, one of Leibniz's successors as royal librarian and historiographer of the House of Brunswick. Scheidt divided the manuscript into forty-eight short chapters, gave them titles, and added the illustrations that Leibniz had planned for his text. These were executed by Nicolaus Seelander, the Hanover court library engraver who had worked for Leibniz in illustrating his *Origins of the Guelfs*.

The illustration clearly belongs with the original text. It was found among Leibniz's manuscript pages and is mentioned in chapter 35, under the heading "Of unicorn horn and the monstrous animal dug up at Quedlinburg." Leibniz explained that the description of the animal was taken from Otto von Guericke, the burgomaster of Magdeburg and the well-known inventor of the vacuum pump. In von Guericke's Latin paper describing his vacuum experiments,[13] this distinguished scientist related the discovery, "in the year sixty-three of this century" (1663), in a quarry on Mount Zeunikenberg in the Harz Mountains, of the "skeleton of a unicorn, the hind part of its body being lowered, and its head raised up and back."[14] Leibniz then went on to repeat von Guericke's description almost word for word.

> Guericke . . . reports the discovery of the skeleton of a one-horned animal. As is usual with such brutes, its posterior parts were very low and its head raised. Its forehead bore a horn nearly five ells long, as thick as a man's thigh but gradually tapering. Because of the ignorance and carelessness of the diggers, the skeleton was broken and extracted in pieces. However, the horn, which was attached to the head, several ribs, and the backbone were brought to the abbess of the town.[15]

This leaves a few questions unanswered. Did Leibniz reproduce an existing drawing? The account of the unicorn's discovery is not illustrated in von Guericke's book, which has quite a few engravings. Did the Magdeburg burgomaster send Leibniz a drawing, which he simply slipped into the sheets of his manuscript? Or did Leibniz try to draw the animal by faithfully following von Guericke's description of its remains? This doesn't seem very likely. The drawing published in *Protogaea* is hardly true to life, but it includes fragments of vertebrae and teeth that resemble the remains of fossil animals readily identifiable today. In 1925 the Austrian paleontologist Othenio Abel[16] recognized them as the combined remains of a rhinoceros and a mammoth. The four teeth in the animal's mouth certainly look like mammoth molars. Also, the scapula is that of a mammoth, as are the dorsal vertebrae, though the spinal column is reversed; curiously, the first cervical vertebra has been placed at the base of the tail. And what are we to make of the "horn" stuck on this bizarre animal's forehead? (Von Guericke explained that when it was discovered, it was still adhering to a fragment of bone.) Could it be the horn of a woolly rhinoceros? Such horn, which is made of keratin, only becomes fossilized under very unusual circumstances (among animals preserved by ice in Siberia, for example). Nor could it be the rostrum of a narwhal, which is spiraled. But it could well be the fossilized straight tusk of a young mammoth, with the jawbone still adhering to the base of the tooth.

Today the *Protogaea* engraving is often cited as the first attempt in the history of paleontology to reconstruct a fossil mammal—but what a bizarre reconstruction it is! Though merely picking out an image isn't enough to understand it; and it must be placed within the narrative that gives it meaning. *Protogaea*, which was probably written between 1691 and 1693, should be seen in relation to Leibniz's other works[17] and also to the scientific issues of his day. The period from 1680 to 1720 was marked in France, England, Italy,

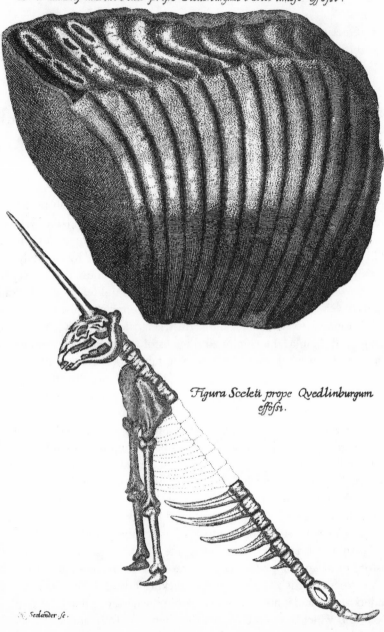

Dens animalÿ marini Tidæ prope Stederburgum e colle limoso effossi . *Tab. XII.*

Figura Sceletí prope Qvedlinburgum effossi .

R. Seelander sc.

"Tooth of a marine animal" and "Skeleton of a fossil unicorn" in Leibniz's *Protogaea*. The tooth is a mammoth's, and the skeleton probably consists of mixed mammoth and rhinoceros bones. Engravings by N. Seelander. (Bibliothèque Nationale de France photo, Paris.)

and Germany by a lively ferment of ideas and theories about fossil objects and the formation of the earth.

In 1691 Leibniz was forty-five years old. He had come to Hanover in 1676 as the librarian and counselor of Duke Johann Friedrich of Hanover and after the latter's death in 1679, that of Duke Ernst Augustus. In 1680 he proposed to write a history of the princely House of Brunswick-Lüneburg. Leibniz saw this as "the opportunity to objectively establish Germany's rights through genealogy"[18] because in his eyes, history was the foundation of right. The task was officially assigned to him in 1685. In preparation for this monumental work—which he would drag "like a ball and chain" his entire life—Leibniz spent three years (from 1687 to 1690) crisscrossing Germany and Italy, checking sources in the archives and libraries in Vienna, Venice, and Rome. This research resulted in the publication of a number of treatises on law and, between 1701 and 1711, documents relating to the House of Hanover-Brunswick, though the *Annales Brunswickenses* would remain unfinished. In his approach, Leibniz's method was that of a true historian,[19] who tries to collect firsthand accounts and primary source documents in order to classify, compare, and discuss them, examine in detail the causes of events, and trace them back to their origins.

As its title says, *Protogaea* deals with "the original aspect of the earth," and the examination of the vestiges of its oldest history in the "monuments of nature" had been planned as the preface to this wide-ranging work. Toward the end of his life,[20] Leibniz intended to use it as an introduction to the history of the House of Hanover, as a "dissertation on the most ancient state of these regions, which can be reconstructed from the vestiges of nature from the time before history" (*Dissertatio de Antiquissimo harum regionum statu, qui ante Historicos ex naturae vestigiis haberi potest*). In writing this text, Leibniz adopted the critical approach of a historian. For his "theory of the earth," he collected the surviving "documents" of this history in the form of minerals and fossils: he had people send him glossopetrae and shells, gathered bones buried in caves, and collected remains.

In listing and describing the objects found in the ground and the caves of Hanover province, Leibniz's approach was less that of a collector or an enlightened amateur than that of a historian. And Germany's history, in this very ancient past revealed by the country's present relief and the objects buried in its soil, is part of the creation and history of the earth and subsequent events—a past that can be deciphered and told. It links the history of present-day humankind to that of the most ancient times on earth. That distance is increasing, however, and the gap widening between the present and that origin, of which we can glimpse only fragments. So it was a matter of collecting these scattered clues that point to the most ancient past of the earth.

The original aim behind *Protogaea* was to write the history of a region, but also what Leibniz called a "natural geography." "We live in the most remarkable and most mineral-rich part of lower Germany," he wrote in the book's opening chapter. This history expands into a much broader one, that of the entire world. Since all history is that of the Creation of man and the earth by God, there is no discontinuity between human history and the history of the earth: "The nature of things fills all the gaps in history for us, and in turn, our history does nature the favor of perpetuating for posterity the knowledge of her notable works, which we are given to contemplate."[21]

To write the history of nature means to give future generations the signs that extend and explain those which nature herself gives us. The last chapter of *Protogaea* lists the "various layers of earth observed in Amsterdam during the digging of a well." It is a veritable "stratigraphy," which establishes the continuous succession of strata from man's present-day life and cultures ("garden earth, seven feet") down to the "sand into which the pilings supporting Amsterdam's houses were driven, ten feet," and far beyond, down to "sand dotted with marine shells," vestiges of diluvial time. Using this vertical series, it is possible to reconstruct the succession of events that took place during the formation of the earth:

> It is likely that the layer in which the shells were found, at a depth of more than a hundred feet, was the ocean floor. Repeated floods and all that they carried with them left the clay and sandbars we have just listed. The sediments [*terrae sedimenta*] were formed in the periods between floods. Driven back, the sea retreated for a while, but returned to assert its rights, burst the dikes and flooded the land, drowning the forests whose remains miners are uncovering today.[22]

Leibniz initially encountered these "documents" and "proofs" of the earth's history not as a scientist, but as a practical man. Between 1679 and 1684, he worked as an adviser and engineer at the Harz mines, which were about sixty miles from Hanover and produced lead, silver, copper, and zinc. Leibniz even designed a number of machines to improve mining operations: windmills for ventilation and devices to pump water from the shafts and bring up ore.[23] Leibniz's knowledge of mining was fundamental to the gestation of *Protogaea*. It turned knowledge about the earth, its history, and the formation of rocks, minerals, and fossil objects into knowledge that was directly useful in everyday human life. It partly explains Leibniz's frequent references in his text to techniques, experimentation, and craft.[24] So his thinking about the formation and very ancient history of the world was linked to concrete, practical preoccupations regarding one nation's history, but also to concerns with the underpinnings of law, economics, and industry. Leibniz focused his study on his country, mentioning German towns and locales, including the Harz mines, the Baumann Cave, and notable natural formations.

Germany's extensive mining activity had sparked the study of fossils since the sixteenth century, as shown by the work of the mineralogist Georgius Agricola.[25] The late seventeenth century was extraordinarily rich in discoveries, inquiries, and speculation about the origin and nature of fossils and the history of the earth. "Systems" had been proliferating for several decades, and Leibniz's culture in this regard[26] is remarkably "European." It drew not only on the works of such German mineralogists and naturalists as Agricola, Kircher, Becher, Lachmann, and Kentmann, but also on those of Englishmen, like Ray, Burnet, and Woodward, and Italians, like Scilla.[27]

But interest in the theory of the earth also was of a speculative nature. Toward the end of the 1680s, Leibniz had read and annotated Descartes's *Principia philosophiae*, which contain—in books 3 and 4—the Cartesian view of the formation of the earth.

In *Le Système du monde* (written in 1633, but published only posthumously, in 1664) and then in the third and fourth parts of *Principles of Philosophy*, Descartes first tried to explain the formation of the earth in terms of physical ("mechanical") laws by relying on "easily understandable and very simple principles, in accordance with which we can see clearly that the stars and the earth, and in fact the entire visible world, could have been created as if from a few seeds." As Descartes told it, stars and their vortices were formed from matter swirling in circular motions in a universe without vacuum. The earth was a sun that had become dark. Within it, several concentric layers became differentiated, the most superficial of which, being less massive, cracked and collapsed. This collapse gave rise to today's oceans and mountains, which are, as it were, the ruined remains of the original planet.

This formulation was a decisive event in the history of ideas about the formation of the earth, in that it attempted an explanation in terms of rational principles and mechanical laws. But it also had an ambiguous status, because it paralleled the Genesis account without contradicting it. Descartes called his history of the earth a *fiction*, saying that the causes his "hypothesis" relies on are both "false" from the point of view of revealed religion, but logically true with respect to natural laws. "Their falseness," he added, "does not prevent what may be deduced from them from being true. . . . The laws of nature [are] such that even if we postulated the chaos that poets speak of, that is, a vast state of confusion in every part of the universe, one could demonstrate that by these laws' action, this confusion must gradually return to the order that is seen in the world today."

Descartes described the stages of the formation of the universe and the earth as an irreversible series of events leading from chaos to the world we see today. But the Cartesian "fiction" about the formation of the earth is not, properly speaking, a *history*; it is based not in reality, only in logic. The history the philosopher re-creates, starting with the present aspect of the world and the mechanical laws of motion that he believes he can establish, essentially relies on deductive reasoning. But thinking about history by using logical reasoning alone does not make one a historian. One must also account for the accidental succession of facts through the detailed study of events.[28] Descartes was not really working as a historian; he never attempted to research the "archives of the earth" for the documents or traces bearing witness to that past.

In his *Protogaea*, Leibniz both built on and criticized Descartes's approach. For Leibniz, it was not a matter of establishing logical "principles" according to which the world could constitute itself, even out of "chaos." Such an approach would risk eliminating the very notion of God in his role in the Creation. One could not imagine that chaos was either at the origin or a part of the history of the world without falling into materialism and irreligion. "God does not do anything without a reason" (*Deus incondita non molitur*) is *Protogaea*'s point of departure. God chose to create our world and its order as the best of all possible worlds. In this world, beings coexisted by virtue of a "preestablished harmony." Leibniz believed in the preformation of living beings (the embryo contains the preformed history of the individual and its descendants), and he also believed in a "preformation" of the transformation of the world. For Leibniz, history was but the development of this divine project.

This was a linear history, and a spectator seeing only a part of it couldn't always understand its inherent global necessity. To a more global vision (which would be that of God), the destructive events, accidents, and catastrophes that afflict the world are not negative; they acquire meaning and purpose in relation to a project of the whole.[29] Thus the disorder we perceive in the world today is only apparent and will lead to order and perfection.[30] This is why the narrative given in *Protogaea* was not presented as a "fiction," but as a true history. And far from being merely "abstract," the Leibnizian vision of nature is at once logical and historical, and relies on concrete, observable proofs.

Protogaea describes the history of the earth this way: The globe, originally consisting of molten matter, hardened as it gradually cooled, eventually becoming solid. The results of this fusion were "vitreous" matters, bare rocks, and flints, which constitute "the great bones of the earth," as well as sand, made up of "small translucent stones." The cooling of the earth gave rise to its relief—mountains and valleys—and "enormous bubbles," which created cavities containing air or water. "Aqueous vapors" above the cooled earth condensed into water, leaching out salts to form oceans. Parts of the earth's crust collapsed under the weight of the water, causing great floods that "left large quantities of sediments in certain places," which hardened when the water retreated. These primordial events may have been followed by other, lesser upheavals, such as earthquakes, localized floods, and volcanic eruptions.

This temporal succession does not quite match the biblical account. Leibniz included the Flood in his account, but less as a miracle and a punishment inflicted on humankind by God than as an explicable physical event—as one episode among others. For that matter, Leibniz posited several "floods" that might have occurred during the formation of the earth.

The history of the earth occupies only the first chapters of *Protogaea*, but long sections of the book are devoted to the question of fossils. In his thinking about fossils and the formation of the earth, Leibniz was greatly inspired by the work of the Danish (later Florentine by adoption) naturalist Nicolaus Steno, whom he probably met when the latter visited Hanover in 1678.[31] The work of this naturalist, who was both profoundly religious (he was born Protestant, converted to Catholicism, and eventually became a bishop) and a rationalist (he was a disciple of Descartes), was an essential source for *Protogaea*.[32]

Steno was born in Copenhagen in 1638, studied medicine and anatomy in Denmark and Amsterdam, then joined the court of Ferdinand II de' Medici in October 1666. That year a huge shark was caught at Leghorn, and Steno was able to dissect it. He showed that glossopetrae—which were found in abundance on the island of Malta and since the Renaissance had always been identified as petrified snake tongues—were in fact shark teeth. In his *Dissertation on the Dissection of a Shark's Head* (*Canis Carchariae Dissectum . . .*), a shark's jaw and an isolated tooth appear in the same illustration, proving the organic origin of the glossopetrae. The image was already well-known in Steno's time; in the sixteenth century, it occurred in the works of Conrad Gesner, Colonna, and Mercati. What Steno did was to systematize and rationalize knowledge that at that time existed only as hypotheses. He took a position in favor of the organic origin of fossil objects, a matter that had been debated for more than a century. "Bodies that resemble plants and animals, which are found in the earth, have the same origins as the plants and animals they resemble," he wrote.[33] At the same time, Steno was trying to understand the mechanical principle behind the formation of fossils. If the "petrified tongues of Malta" were once shark teeth, how did they come to be within the layers of the earth, sometimes far from the sea? This inquiry into the origin of fossils would lead to a more general meditation on the formation of "solid bodies

. . . naturally enclosed within another solid." Which is precisely the formulation of the problem that Steno raised in a celebrated text called *De solido intra solidum naturaliter contento dissertationis Prodromus* (1669). "If a solid body is enclosed on all sides within another solid body, of the two solids, the one that hardened first is the one that expresses its surface properties on the second surface."

If a solid has been enclosed within another solid, one must conclude that they were formed not simultaneously, but successively. So fossil shells and fishes found in the earth—solids naturally enclosed within a solid—existed as such before their inclusion, and therefore were not formed in the earth, as the Renaissance encyclopedists believed. The spatial relationship of inclusion expressed a temporal relationship of succession. This simple, rational ("geometric") principle also accounted for the successive formation of layers of the earth in function of the debris they contain. The last part of *Prodromus* examines how the relief of Tuscany was formed, as a series of collapses and sedimentations, among which the biblical Deluge appears as an essential episode.

Shark's head illustrating Steno's demonstration of the organic origin of glossopetrae. The baroque shark is actually from a sixteenth-century engraving by the Italian naturalist Michele Mercati, which was borrowed by Steno, and later published in Leibniz's *Protogaea*. (Bibliothèque Nationale de France photo, Paris.)

It is precisely this rule of inclusion that Leibniz used as a principle for his explanation of the origin and formation of the earth. Where Steno studied the relief of Tuscany, Leibniz attempted to create a general theory of the formation of the globe, starting with the "natural geography" of the Hanover region.

> It is certain that if the globe had originally been *liquid*, it should have had a smooth surface. And it is consistent with the general laws of bodies that, as liquids thicken, they produce solids. This is shown by the layers and nodules deposited in the interstices of rocks, such as mineral veins and gems.
> Moreover, one occasionally finds debris from the past— plants, animals, and handcrafted objects (*arte factorum*)—enveloped in a sheath of stone.
> Consequently, this solid envelope must be of recent formation, and it must have been originally in a fluid state.[34]

The fossil remains of marine animals and terrestrial quadrupeds bore witness to this history. It was because the earthly globe had originally been liquid that one found, as Steno explained, "solid bodies enclosed in solid matter," minerals and gemstones locked in rocks, "*debris from the past*, plants, animals, and works of art now enveloped in a sheath of stone." The presence of "solid bodies" *within* a solid milieu could only be explained if one supposed a chronological succession of events. The petrifaction of living beings and their insertion into the layers of the earth necessarily involved time.

Through his "historical" perspective, Leibniz thus freed himself from the hermetic tradition and from alchemy, of which Kircher's 1665 *Mundus subterraneus* gave a belated and remarkably complete picture. Leibniz himself had once admired Kircher and believed in "figured stones." In a manuscript— undated, but earlier than *Protogaea* and most likely written before he met Steno in 1678—Leibniz wrote:

> I find it hard to believe that the bones that one sometimes finds in fields, or which can be found by digging in the earth, are the remains of actual giants; likewise, that the stones of Malta, which are commonly called snake tongues, are parts of fishes; or that the shells that one sees at some distance from the sea are any certain indication that the sea covered those areas and had left the shells behind when it withdrew, and that they were later petrified.

If this were the case, the earth would have to be much older than is reported in Holy Writ. But I do not want to dwell on that; we are concerned here with applying natural reasoning. I believe that the shapes of these animal bones and shells are often merely sports of nature, which have been formed apart, without coming from the animals. It is well-known that stones grow and assume a thousand strange forms, as shown by the stone figures collected by the reverend father Kircher in his *Subterranean World*.[35]

At the time, Leibniz firmly believed in "sports of nature" and "petrifications" through the action of a plastic faculty [*plastica facultate*] of the earth. He refused to accept that fossil shells or bones could have been deposited and petrified "several thousand years ago," probably as much for logical reasons as belief in religious dogma. "If this were the case, the earth would have to be much older than is set down in Holy Writ."

The age of the earth is not touched on in *Protogaea*, but neither is there any more talk of the plastic faculty of the earth or the "growth of stones." To the contrary, Leibniz could hardly find language strong enough to criticize the "deplorable facility" and "wandering imagination" of those who believe in "figured stones," to condemn the vanity and credulity—which he once shared—of "childish tales seriously set forth in works that are almost contemporary, by Kircher, Becher," and other authors who have written about "the marvelous sports of nature and her generative power [*vi formatrice*]."[36]

When they resembled existing living beings, those buried remains could not be the product of "sports of nature" or of accidents, as was formerly believed. Their presence must be related to a *history* of the earth that includes episodes where these living beings in fact could have existed, and others during which they were buried in the layers of the earth where they are found today. Such an approach supposes two things: first, that one can distinguish among "fossil" objects within the earth itself—animate from inanimate, mineral from living; and second, that one can rationally explain the objects buried in the earth.

Leibniz put the question of *resemblance* at the heart of the discussion of the nature of fossils. Among fossil objects, he said, we must first distinguish inanimate objects—for example, crystals, whose geometric shapes and "polygon figures" can be easily explained by the "juxtaposition of their parts." There are also figures "which only a distracted imagination could see in the stones."[37] Some people have discerned historic, mythical, and even religious scenes in stones, making out "Christ and Moses on the walls of the Baumann

Cave, Apollo and the Muses in agate veins, the pope and Luther in Eisleben stone, and the sun, moon, and stars in marble." In those cases, "credulity enhances the rough-hewn work of accident," bringing to light tricks that are "not even [those] of nature, but of the imagination." Finally, there are images that exactly resemble living creatures. Such as the "imprints of fishes in sheets of stone," which are mainly found in the coppery shales at Eisleben, in Saxony. These "forms of fishes" imprinted in the schist are "clearly and exactly rendered, as if an artist has slipped a sheet of engraved metal into the black stone; some call them ichthyomorphs."[38] This last class of objects required a special explanation, distinct from the others. One cannot believe "that organic bodies without antecedents, purpose, or seed [*sine seminiis*] would come into being outside all the laws of nature by who knows what plastic force within the inert matrices of silt and stone."[39]

Natural causes must be invoked to explain the presence of these living beings' remains in the rock. Their occasional abundance in a given site, the diversity of perfectly recognizable species represented, and the fact that some remains were intact and others broken, all argued for a single cause of their being there. "The multiplicity of these images in a single spot is so great that we would more likely suppose a manifest and constant cause than a play of fate or of some 'generative idea,' useless words behind which philosophers hide their ignorance."

It was therefore necessary to go back to the origin of these "images," and to recognize in them the imprint of animals that actually lived. Not that image and imprint are necessarily the same. If the imprint is the real trace of a being that "passed by there," the resemblance that the image bears is not always the sign of a real presence. For supporters of "sports of nature," "ichthyomorphic" (fish-shaped) stones, were—like the other representations in stone of objects, images, or scenes—examples of the "genius of things" and proof of the ability of "nature, that great creator, playfully [to imitate] animal teeth and bones, shells, and snakes." Outside of the living world, the supposed imitation of animal shapes could only be partial and approximate. And when these imitations appeared to be perfect and so precise that one could recognize "the species of the fish at first glance," without there having been any change in proportions or symmetry, this proved that what is involved was not a copy in stone of the animal's appearance, but indeed the imprint of the animal itself.

Thus the explanation of fossils by the accident of a mimetic game of nature stood refuted. "If nature were playing, she would play with more freedom, and would not limit herself to expressing so exactly the tiniest traits of the original while preserving their dimensions so exactly, which is even more remarkable," explained Fontenelle, in summarizing a paper that Leibniz sent to the French Academy of Sciences in 1706;[40] "chance never exactly duplicates nature."[41] Taking a "naturalist" approach, one can recognize in these fossils the fishes' general anatomy and even distinguish between different species: "I have in hand fragments in which images of a mullet, a perch, and a bleak are imprinted in the rock. . . . I have also seen saltwater fishes, like ray, herring, and lamprey, with the latter sometimes lying beneath the herring."[42] In this respect what is convincing is the feeling of life that emerges from these imprints' shape and precisely detailed character and the positions of the animals. "They brought from the quarry an enormous pike, its body bent and mouth open, as if it had been taken by surprise while still alive, and resisted the petrifying force."

Identifying fossils as the remains of living creatures—or rather creatures that were once alive—is to consider them no longer merely as object or images, but as signs that point to a past. In short, fossils constitute the language of nature, which in the present speaks of the past. This concept was similar to the etymological approach that Leibniz took in his work on languages.[43] But where language is a human creation and has meaning independently of etymology, to understand the language of fossils one must question the very means by which they were created. The model of the technical and craftsmanlike activity plays an essential role, wrote Leibniz, in an approach that aims to understand and know the operations of nature. "We feel that one who carefully compares the products of nature drawn from the bosom of the earth with the products of laboratories . . . will have accomplished an important task, for the striking links between the products of nature and those of art will leap to our eyes."[44]

Not only is "the formation of minerals explained by chemistry," but some substances can be produced in chemistry laboratories in much the way they are in nature, such as cinnabar, zinc, and arsenic. Explaining the "petrification" of living beings supposes that one takes into account the passage of fluid materials that turns them into solids; this hardening process can be explained through the example of human craftsmanship.

Human behavior resembles divine action when it uses the properties of natural elements in its own productions. Imitating the ways of nature is a way of knowing her: "Finding a way to produce things is already a great step toward understanding them."[45] Technical activity is one of the paths of science. Leibniz henceforth rejected the alchemical interpretation of the "vegetation" of stones and the growth of minerals within the earth. In his explanation of fossils, he refuted the hypothesis of an accidental *mimesis* of nature and drew a parallel between "nature's hidden operations" and the "visible works of men." He cited the technique of firing clay pots as an example of the hardening of earth that was once liquid and referred to the goldsmith's art to explain the process of fossilization:

Having enveloped a spider or some other insect in an appropriate substance while leaving a slender tube open, they harden this substance by fire. Then, by pouring mercury in, they drive the animal's ashes out through the small opening, replacing them by then pouring molten silver in through the same hole. Finally, on breaking the mold, they reveal a silver animal with its entire apparatus of legs, antennae, and fibrils, astonishingly true to life.[46]

Through his "art," man therefore perfectly imitates life down to its finest detail: the silver insect, "with its entire apparatus of legs, antennae, and fibrils," in this way is very much like the fish whose detailed, engraving-like image is imprinted on the rock. In *Protogaea*, Leibniz strongly insisted on the human ability to imitate the actions of nature by using her laws (to the point of producing a copy that is "astonishingly true to life"). This simple statement leads to the idea of an *experimentation* through which it is possible to know the ways of nature by borrowing them, and to discover the logic of her history through man's present action.

As the seventeenth century ended, the organic origins of fossil remains became widely accepted by the scientific community. Leibniz himself took the lead, moving from a belief in "figured stones" to recognizing animal remains as proof of the earth's history. In this, he was the heir to an entire tradition that is French (Palissy), Italian (from Leonardo da Vinci to Scilla and Steno), and English (Hooke, Burnet, and Woodward). It is true that in 1721 his disciple Louis Bourguet, in his *Lettres philosophiques sur la formation des sels et des crystaux* (Philosophical letters on the formation of salts and crystals), still adopted a polemical tone in refuting the "sports of nature" thesis, and in 1749 Buffon had to revive the entire debate against Voltaire. But those polemics, while spectacular, were only rearguard skirmishes. For the great majority of European naturalists, the matter of the identification of fossils was settled by 1720.

Still, one problem remained, that of fossils that resemble no living being known in nature today. Must one accept, as Bernard Palissy did a century earlier, the idea of certain animal species having disappeared? Could ammonites—Leibniz illustrated several varieties of these "horns of Ammon" in *Protogaea*—be "lost species"? He found that unthinkable and imagined that they must exist in distant seas or at depths yet unexplored—perhaps on the ocean floor or in the seas of the Indies. It was impossible for Leibniz to conceive of extinct species, because for him, the world had been created perfect. The "scale of beings" was established once and for all as an immutable hierarchy.

On several occasions, Leibniz seemed to briefly consider the possibility of living creatures having changed during the course of the earth's history according to the circumstances and the milieu in which they lived, but only quickly to reject it as theologically dangerous. "I am not ignorant of the fact that some push the boldness of their conjectures to the point of thinking that when the ocean covered everything, the animals that then populated the earth were aquatic, and that they became amphibian as the waters receded, and that their posterity wound up abandoning their original habitat." He then cautiously added: "But quite aside from the fact that these opinions are in opposition to Holy Writ, from which we should not stray, the hypothesis on its own merits presents inextricable difficulties." That thesis was of a clearly antireligious and materialistic character, particularly when it was picked up and systematized in the same period by such libertines as Benoît de Maillet in his *Telliamed*.[47]

The problem of the presence of quadruped bones in the Harz caves is the same raised by fossil fishes and shells. If one admits the existence of "intact bones of marine monsters and other animals of an unknown world" in the layers of the Baumann and Scharzfeld caves, this can only consist, according to Leibniz, in "skulls and lower jaws of great animals in which the teeth are still implanted in such a way as to eliminate this ridiculous misconception of teeth born of the earth."[48] As to "those massive teeth said to belong to the elephants known to the Russians under the name of *mammoths*, and which [the Dutch traveler Nicolaas] Witsen suggests are the bones of large animals,"[49] were these marine animals or elephants? Leibniz didn't deny that "real elephant bones" were being found, and he displayed a lively interest in the discovery in 1695 of the skeleton of a "petrified elephant"[50] found the previous year in the village of Tonna in Upper Saxony—a discovery that would spark new debates about the nature of fossil objects. Leibniz was asked for his views of the specimen. "In Tonna, near Gotha in Thuringia, several parts of a skeleton were found which to all appearances were that of an elephant," he wrote to one of his correspondents in June 1696. "Several doctors there claimed that it was a product of the earth, *lusus naturae*. When I was consulted, I said I had no doubt it was *ex regno animali*, and that if it was not an elephant, it was some similar animal."[51]

In *Protogaea*, Leibniz recognized the existence of the fossil remains of elephants. "I have seen," he said, "what are certainly teeth, a fragment of tibia, and other bones excavated from the Scharzfeld Cave, which cannot be attributed to any animal other than an elephant." But where did these elephants come from? Leibniz suggested two hypotheses. The first was that "these animals [had been] much more widespread on the surface of the globe than they are today," which could be explained by the fact that "the nature of the soil [had] changed." The second was that "their bodies [had been] carried far from their native lands by the violence of waters." Of these two hypotheses, Leibniz seemed to favor that of a universal flood, corresponding to the biblical Deluge. "I am inclined to believe," he wrote, "that this accumulation [of bones in caves] is the result of a vast flood, following which the waters, in seeking a passage toward the interior of the earth through narrow fissures and the secret exits of caves, left the bodies they were carrying in the vestibule, as it were."[52]

Which brings us back to the "unicorn" that first caught our eye. According to Leibniz, the caves of Germany also contain the remains of "a unicorn quadruped of the size of a horse." Is this a return to medieval myths and Renaissance beliefs? Leibniz knew perfectly well that the famous "unicorn horns, once one of the most curious and rarest ornaments of natural history cabinets, and today the subject of the people's admiration," had been recognized as twisted narwhal teeth since the Dane Thomas Bartholin identified them. Nor did he subscribe to the old belief that saw unicorn horns in any number of whitish fossil objects. Still, he thought these remains were those of a living animal with a single horn, and he drew on solid evidence in support of this belief. The discovery of this "unicorn fossil" was presented as a "fact" supported by the testimony of the very respectable statesman and scientist Otto von Guericke. The discovery was authenticated by a great deal of evidence. The visual evidence ("I saw") is joined by the equally reliable—in Leibniz's eyes—testimony of people worthy of belief, that of two Portuguese travelers, "Hieronymus Lupo and Balthasar Tellesio," who claimed to have encountered a unicorn in Abyssinia.

An ultimate proof lay in the work of the anatomical reconstruction and illustration that accompanies his text—the first time a terrestrial vertebrate was reconstructed from collected fossil bones. Leibniz may have used Guericke's reconstruction of the animal's skeleton, knowing that ignorant and careless "diggers" had "broken" the skeleton and "brought it out in pieces," with only the "horn attached to the head, several ribs, and the backbone" remaining. While barely taking into account the fact that the skeleton was incomplete (though he noted it), Leibniz presented the astonishing picture of a two-legged animal with the head of a horse and the molars of an elephant. In this representation, where traditional mythical imagery joins a supposedly rational approach, Leibniz's "unicorn" is a chimera in the true sense of the word.

Leibniz's method in these pages illustrates in more ways than one the characteristics of natural history at the end of the seventeenth century. The anatomical approach is joined by hearsay knowledge of elephants and Siberian mammoths, belief in accounts whose reliability depends on the quality of the witnesses and references to Scripture. "The fossil unicorn found in our country" proves the transport of their petrified remains, which can be explained, like that of fossil elephants, by the action of the waters of floods once covering continents that are now dry.

In fact, all the elements needed to understand the Quedlinburg discovery were already in place. Leibniz no longer believed in fossils existing sui generis in the earth, nor in the medieval myth of unicorn horn, whose miraculous properties had been so long proclaimed (though lately less so) by collectors of curiosities and wonders. He could tell a narwhal rostrum from a land animal's tusk. A few paragraphs earlier,[53] he wrote about elephants and mammoths, and, a few lines higher, remarked that the remains of the Quedlinburg animal were incomplete. But Leibniz had probably never seen an elephant. And the "Russian mammoth," which had become known only recently from the accounts of European travelers to Siberia, remained an animal far more mythical than the unicorn, long familiar to European eyes and dreams. Since the Middle Ages, the unicorn lived in stories, was the subject of tapestries, and graced cabinets of wonders. If Leibniz had wanted the skeleton of this improbable two-legged beast, with its huge molars and single forehead horn, to appear among the illustrations of his book—despite the requirements of rationality shown in his thoughtful contemplation of fossils and the history of the earth—it is because at the dawn of the Enlightenment, the fabulous bestiary still remained a credible source for zoology.

Identifying an Elephant:
The Russian Mamont,
the Elephant, and the Flood

When the spring thaw comes to the frozen plains of eastern Siberia, huge curved tusks sometimes appear on the surface of the ground. They are usually found singly, but sometimes as part of a skull or an entire skeleton. They are the remains of prehistoric animals that lived in these regions, at the distant time when our ancestors were still chipping stones into tools. Carried by streams and watercourses, the bones pile up along the banks or mouths of great rivers. In some parts of the Russian plain and on the islands to the extreme north of Siberia, they literally carpet the ground.

Once in a while, the thaw reveals the entire body of an animal buried for millennia in the frozen ground, its flesh and bones virtually intact. The Siberian permafrost has preserved the flesh and organs of these animals, extinct for thousands of years; sometimes the flesh is still edible, and the natives occasionally eat it. For centuries they have used fossil ivory to make jewelry and useful objects of all kinds. But they are not interested in these remains only as a source of food or trade. At least until the middle of the twentieth century, this mysterious animal, which they call "mammoth" and have never seen alive, played an important role in their sacred bestiary and mythology.[1]

The Yakuts of Siberia imagined the mammoth to be a kind of giant rat that flees sunlight and lives underground, where it tunnels along with its enormous horns, making the earth tremble and raising hills on its passage. When it emerges into the light, it dies, so its remains are found on the riverbanks or near estuaries. According to some Yakut tales, the mammoth is the spirit and the ruler of waters. In winter it breaks the river ice with its horns. The Evenki, Seljuk, Ostyak, and Mansi consider the mammoth to be aquatic; other Siberian peoples represent it in the shape of a gigantic bird.

Among these peoples, the image of the mammoth is an intimate part of their belief in shamanism. Some carved pendants on their shamans' costumes,

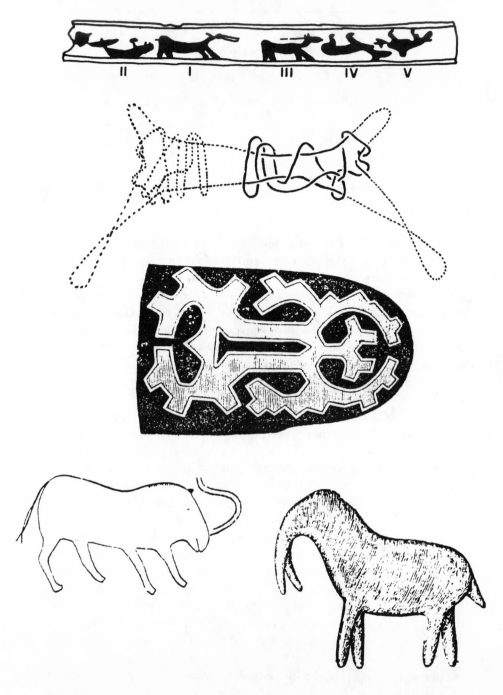

Schematic representations of the mammoth. *From top to bottom,* in Alaska: engravings on walrus ivory; cat's cradle pattern, per André Leroi-Gourhan (1935). In Siberia: Khants embroidery pattern; Yugadir and Khants drawing and pendant, per Ivanov (1949).

which include real and fantastical animals, are mammoth shaped. The "horned animal" that lives underground joins the spirits of subterranean beings, over which shamans wield their power.

On the other side of the Bering Strait, the Eskimos of Alaska also have legends, myths, and rituals concerning the animal of the snows. In pictures etched on walrus ivory, one can see strange quadrupeds with long horns and short, stocky legs, which look like ibex when their "horns" point backward, or reindeer when they point forward. But there are no ibex in Alaska, and these figures, which some ethnologists consider mythical goatlike animals, are probably firsthand renderings of mammoths, based on the shape of the animal's body, deformed by the ice. For these people, too, the mammoth belongs to the realm of underground beings and the mythology of the world of the dead. It is the basis of representations of all kinds, such as the cat's cradle games in which a stretched string creates the silhouette of an animal with an abdomen and massive legs shaped like a mammoth frozen in the ice.

Mammoth fossils have found their way into many kinds of representations among the people of the far north: images, fables, tales, stories, rituals and games, mythical representations, and religious beliefs. From Siberia to China and Manchuria, from the Bering Strait to Baffin Island, objects and images circulate and traditions and tales are transmitted that give these animals from another age a place in daily human life.

The word "mammoth" appears for the first time in Western literature in 1692, in *North- and East Tartary* by the Dutchman Nicolaas Witsen.[2] When he traveled through northern Russia, Witsen was shown the bones and giant teeth that are found on the banks of Siberian rivers. The Russian settlers in Siberia called these remains *mamontekost*, literally "mammoth bone," and the animal itself *mamont* or *mamot*. Where does this word come from? Some authors mention a confusion with the name of Saint Mamas, who is revered by the Russians. The German naturalist Peter Simon Pallas suggests a double Finno-Ugric root: *ma*, meaning "earth," and *mut* or *muit* in Estonian, meaning "mole." An "earth mole"? That seems a bit redundant, and Estonia is far from Siberia, but this origin fits the mammoth's mythical significance to the Siberians: an animal that lives underground and occasionally emerges near rivers. A second root Pallas considered is the word *mama*, which he thought means "the earth" in the Tartar dialect; oddly enough, the word doesn't appear in any dictionary.

Another etymological tradition suggests that "mammoth" comes from "Behemoth," the monstrous beast described in the Bible in Job 40:15–19:

> Look now at the Behemoth,
> which I made along with you;

he eats grass like an ox.
See now, his strength in his hips,
and his power in his stomach muscles.
He moves his tail like a cedar;
the sinews of his thighs are tightly knit.
His bones are like beams of bronze,
his ribs like bars of iron.
He is the first of the ways of God.

This etymology is mentioned by Father Philippe Avril, a French Jesuit who traveled through China in 1685. It may have been brought to Siberia by Arab traders as early as the tenth or eleventh century and was mentioned by Russian explorers in the early eighteenth century. This underscores the very strong connection between natural history and theology at the time.

Whatever the origin of its name, the mysterious "Russian *mamont*" would spark a good number of questions and controversies throughout the eighteenth century.

"By order and all-powerful will of his majesty the Czar, I have been ordered to travel to Tobolsk and to bring back . . . all the curiosities to be found in the region of Siberia, including objects of antiquity, pagan idols, large mammoth bones, and ancient Kalmuk and Tartar manuscripts. As soon as I have gathered them, I can send you drawings and descriptions of them." So wrote the naturalist Daniel Gottlieb Messerschmidt to Prince Tcherkaski,[3] the governor of Siberia, in 1720.

Peter the Great, the czar of the new Russia, decided in 1714 to assemble a cabinet of curiosities at St. Petersburg modeled after those of the princes of western Europe. In 1718 he issued a special order to collect "various extraordinary things" for his *Kunstkammer*. Here are the terms of this ukase: "If anyone should find in the earth or in the water an ancient object, an unusual rock, human, animal, fish, or bird bones that are similar to the ones known today, or are exceptionally large or small, he should bring it back, and will be paid according to the recovered object's worth."

It was also at Peter the Great's initiative that the St. Petersburg Academy of Sciences was founded in 1725, which played an active role in organizing scientific exploration expeditions in Russian territory. Their object was to increase knowledge of that immense country's geography, its farthest regions and their inhabitants, to study its flora and fauna, and to inventory its underground riches. As early as 1722, Peter the Great specifically ordered research on the mysterious animal the Siberian natives called mammoth. If anyone found its "horns," they were instructed to search the nearby area very carefully for the entire body of the animal and to send it to St. Petersburg. The expeditions multiplied and remains were collected. In 1723 "superintendent Nasar Kolechov brought the head of an extraordinary animal to Irkutsk: it was three and a half feet long, two feet high, with two horns and a molar tooth. These objects were displayed in the imperial natural history cabinet and described as mammoth bones."[4] But what was this animal, which no one had seen alive and whose bones were being found everywhere, including its "horns" and sometimes, it was said, bloody flesh?

The mammoth excited fear and curiosity, but its wonderful "horns" also sparked considerable greed. Siberian natives had long made use of fossil ivory, and from time immemorial, China had imported ivory from Siberia. Its trade with central Asian and Arab merchants had been known since the ninth century.[5] After the Russians conquered Siberia in 1579, cossacks, hunters, sailors, naturalists, and explorers set out to discover the great Siberian north.

> Cossacks and hunters made the first discoveries. They were the
> first to sail their primitive boats called "kotcha" or "chitik"
> along the north coast of Siberia from the estuary of the Lena
> River in the west to the Pacific Ocean in the east. History has
> preserved some of their names: Elisei Buza, who between 1631
> and 1642 sailed from the Lena River's mouth as far east as that
> of the Yana; Mikhail Ivanov, who reached the mouth of the In-
> digirka; Stadukhin, who reached Kolyma in 1644; and the well-
> known Dejnioff, who had the honor of opening the sea route
> from Kolyma to the river Anadyra on the north Pacific Ocean,
> in 1648. One should also mention Buldakov, Nikifor Malguin,
> Yakov Viatka, Permiakov, etc. What gave them the courage
> to undertake these expeditions was the hope of finding the
> "golden fleece"—sable furs. When that resource ran out, a
> promise of new riches appeared: mammoth tusks.[6]

The "holy terror" that mammoth remains inspired long limited the Russian cossacks' interest in fossil ivory. But in the early eighteenth century, Peter the Great's repeated ukases made the animal better known by making it less fantastic and also created the possibility of a real trade in fossil ivory with the countries of Europe. "Under the pretext of seeking mammoth bones, the cossacks of Yakutsk undertook great voyages [in Siberia]," reported the German traveler Johann Georg Gmelin.

> Where a single horse apiece would have sufficed, they were
> given five or six, which they loaded with their merchandise.

Encouraged by this support, they all wanted to go looking for bones. Before then, a mammoth skeleton, or even a reputed one, was a holy thing that no one would dare touch. The cossacks were even afraid to look at these sinister remains from a distance. But when the emperor demanded that they do so, they feared it would be *lèse-majesté* not to follow his orders.[7]

During the eighteenth century, new mammoth bone sites were found in the islands along the northern Siberian coast. In 1750 a Yakut merchant named Ivan Lyakhov discovered a peninsula covered with bones between Khatangskiy Bay and the Anabar River. Through a Yakut named Eterikan, he learned that even richer islands lay north of Cape Svyatoy Nos ("holy promontory"). One day in 1770, he left with a dogsled to explore two of these islands, which would later be named Big and Little Lyakhov in his honor. A third island, the largest of this group, which he reached by boat in 1773, was named Kotel'nyy, or "kettle," Island because he left a copper kettle there. Lyakhov informed the government of his discoveries and was given the exclusive right to take mammoth ivory and arctic fox skins from there and from any islands he might yet discover. Mammoth bones and ivory have been collected on the Lyakhov Islands ever since. It was there that the paleontologist K. A. Volosovich gathered a complete mammoth skeleton in 1910. After many peregrinations,[8] it wound up in the drawers of the Muséum National d'Histoire Naturelle in Paris, where it was "discovered" in 1957 by paleontologist Yves Coppens. Mounted according to his instructions, it is now on display in the Muséum's paleontology hall.

Europe discovered Russia's fossil ivory around 1725, and that precious material has been exported to European markets ever since. In the mid-nineteenth century, the naturalist Alexander Middendorf estimated that two hundred tons of ivory had been sold during the two previous centuries, equivalent to the ivory from a hundred mammoths every year.

Peter the Great's encouragement spurred both economic conquest and scientific research, and a number of exploratory missions were dispatched to Siberia. One goal of the naturalists who joined those expeditions was to untangle the fabulous from the true in their predecessors' accounts about the mammoth.

The precious "horns" were found in quantity in the northernmost parts of Siberia, especially on the islands and the coasts, or along the great rivers. But many occurred all over the great eastern Russian plain. At the end of the eighteenth century, the naturalist Peter Simon Pallas wrote:

> In all of Asiatic Russia, from the Don to the tip of the Chukchi Peninsula, there is not a river, especially among those that flow

through the plains, that does not have bones of elephants or other animals foreign to those climes along its banks or in its bed. Those bones and marine petrifications are absent from the primitive and schisty mountain uplands, whereas the lower slopes and the great alluvial plains exhibit them wherever they have been eroded by rivers and streams, which proves that one would find as many in the rest of their expanse, given the same means of excavating them.[9]

While journeying through Russia and China in 1692, the German traveler Evert Ysbrants Ides mentioned the discovery of a mammoth head and foot, with flesh attached, preserved in the ice.

A traveler who came to China with me, and who went searching for mammoth bones every year, assured me that he had once found one of those animals' entire head in the frozen earth, with the flesh rotten. Its teeth protruded from the snout like elephant teeth, and he and his companions broke them and several bones of the head off with great difficulty, as well as neck bones, which were as if still stained with blood. Finally, having searched further in the same area of the ground, he found a frozen foot of monstrous size which he brought to the city of Tragan. According to the traveler, this foot was as big around as a large man's waist.[10]

Other eyewitnesses, such as Captain John Philip Muller, told of extraordinary finds: "Muller describes the mammoth as being eight or ten feet high, some eighteen feet long, gray in color, with a long head, wide forehead, and having above its eyes two horns that it can move and cross one over the other. When it walks, it stretches out considerably, but it can also squeeze itself into a small space. Its paws are the size of a bear's."[11]

To Gmelin, who also traveled though Siberia during the first decades of the eighteenth century, these were nothing but fabulous stories. "They said they had found [remains] that were still bloodstained. This tale was brought back by Ysbrants Ides, and Muller and others later repeated it as truth. A fabulous tale always grows in the telling; it was added that these bloody bones were those of an animal named mammont that lived underground in Siberia. It had died buried in an avalanche, which is why its bones were found, still bloody." Concluded Gmelin: "Ysbrants Ides candidly admits that no one had been able to tell him of having seen a living mammoth; there is nothing surprising in that; this animal must be ranked with sirens, phoenixes, and griffins."[12] In his skepticism of the Siberian legends, Gmelin taxed the testimony of those who had seen bloody remains as "credulity": if the animal still existed, he argued, someone would have seen it. If it no longer lived in those regions, one would see only dry bones, not fresh flesh. The fabulous mammoth was certainly generating its share of mysteries.

In 1720 an expedition set out led by a Russian mining engineer named Vassily Nikitich Tatischev (1686–1750). Its mission was to find silver and copper ore in Siberia, in particular in Tungus country. Tatischev was the administrator of Russian mines and governor of the Orenburg region, but he also was a man of the eighteenth century. He had a curious and encyclo-pedic mind and a passion for geography, history, mathematics, topography, archaeology, and languages. As a young soldier, he had fought in Peter the Great's campaigns against Sweden; later he became interested in Russian ar-chaeological remains and in collecting antiquities. During his stay in Siberia, he often sent papers to the St. Petersburg Academy about his discoveries of rare objects, fossils, and other curiosities. He drew up a questionnaire that

Manuscript by Vassily Tatischev recording the Siberian natives' tales of the *mamont* animal of Siberia (ca. 1722). The word "mammoth" is underlined in the title. (St. Petersburg Academy of Sciences archive.)

he sent to various Russian regions to study the country's geography; in it, he mentioned the importance of collecting "subterranean petrifications." In 1725 Tatischev published in a Swedish journal[13] a letter in Latin about the bones of the animals the Russians called *mamont*. It was quoted in a text published in Russian in 1730.[14]

With the greatest precision, Tatischev first described the circumstances surrounding these discoveries: "Fishermen and hunters come across these animals' bones by accident, especially in northern areas and in Tungusy, Yakutia, Daoury, Obdary, Udary, Kondory, and Ugory provinces, near rivers where water overflows their banks and the earth collapses. These bones can then be seen, still stuck in the earth at a depth of 2 to 20 sajens [14 to 140 feet], near the bank, or even in the water."[15]

Tatischev remarked that one does not find the famous ivory "horns" alone, but also skulls, vertebrae, ribs, and tibias, which are not collected, having no value. The "horns" vary in size and quality: "Some are small—not more than an arshin [30 inches] long and a vershok [1.73 inches] thick at the base. Others are almost 4 arshins long and 5 or more vershoks thick. The biggest horns weigh as much as 7 poods [250 pounds]." Their aspect and color also varies; some are very fresh and hard, others already rotten; some are solid and white, other yellow and blackened, wrote Tatischev, drawing on the ivory hunters' already considerable knowledge. "Because these bones are very heavy and curved, it is awkward to carry them to distant settlements, for lack of water routes or pack animals. So they are cut in pieces, which are then brought back for sale to sculptors and carvers and to the merchants who trade with China, where they fetch good prices. A large quantity is shipped to Moscow."[16]

Tatischev also mentioned the legends that the native Siberians told about the mammoth: it was said to be an animal the size of an elephant or a little bigger, black in color, "with two horns on its head which it can move as it pleases, as if they were not solidly anchored." The Siberians add that "this animal always lives underground; it moves from one place to another, clearing its way with its horns. But it can never come out into daylight. If it gets very close to the earth's surface and starts to breathe air, it dies. They imagine that for food, it eats the earth itself. Some also say that it cannot live in the same places as man, and that it lives in deserted areas in order to avoid humans."[17]

How much faith should one put in these statements? Does this animal really exist? Should one reject native legends in favor of more rational explanations, or is there some truth to them? It is true, Tatischev remarked, that the animals' entire bodies had sometimes been collected in a good state of preservation. "Some say that when they pulled the horns out of the mountainside, they saw blood flowing and concluded that the horns belonged to animals that were still alive, or which had only recently died."[18] The remains varied greatly in appearance: rotten bones must come from animals that died a long time ago; fresh and sometimes bloody remains could belong to animals that died more recently. As to the legend that these animals live and walk underground, it is not unbelievable: subterranean animals certainly exist; some of them are blind, and when they come out to the surface, they die. Others, like marmots, skunks, and moles, live both on and in the earth.

When all is said and done, what is the nature of this animal? From the horns' appearance and curvature, they could be those of a bull. Or might they be elephant tusks, as some claim? "It is not possible that elephants could have reached such latitudes without having special fur, because they would have died of cold . . . and if these living animals had actually inhabited these areas, it is impossible for men not to have seen them, if only once. . . . And in this case, how could their bones be buried so deeply in the earth?"[19]

A number of hypotheses had recently been suggested, including the somewhat fantastical one by a Swedish author: "The ancient peoples of Siberia . . . the Ostyaks, Votyaks, Tungus, and others, are the distant descendants of the Israelites ten generations ago. . . . When the Jews crossed the burning deserts, they brought with them many elephants; and when they reached northern countries, those elephants died of cold and have been sinking into the ground since those ancient times."[20] Others affirmed that Alexander the Great led his campaign to Siberia and brought so many elephants that a large number of them were buried in the ground, where they are found today. According to another version, the Chinese brought elephants to these parts during their wars; however, our cautious engineer Tatischev added, historical proof is missing, as all the manuscripts on the subject have disappeared.

Tatischev also noted the explanation according to which these elephant cadavers had been carried by the waters of the Deluge from the warm regions where they live today to Siberia and were then buried in the depths of the ground by earthquakes. This hypothesis, which takes into account the biblical episode of the Flood, was very often mentioned starting early in the eighteenth century, notably by Ides,[21] Gmelin, and Pallas. But if those elephants had been carried by the Flood, why were some of their bones fresh and others rotten or petrified? Tatischev prudently concluded: "If some notions still remain doubtful to this day, and we are not able to put forth more certain ideas, we hope that they will nevertheless stimulate the thinking of enlightened readers, when more of these mammoths will have been gathered and studied, and that light will be cast on this."[22]

Another important 1720 expedition to Siberia that lasted seven years was led by Daniel Gottlieb Messerschmidt, a German doctor, geographer, and naturalist who studied medicinal plants and epidemic illnesses. He under-took to describe the peoples of Siberia, the monuments of antiquity, and the extraordinary things he encountered during the expedition. Seven volumes of journals and unpublished notes from his voyage to Siberia are stored in the archives of the St. Petersburg Academy of Sciences. Messerschmidt, like other travelers to Siberia, soon became aware of the famous *mamontovokost*. In 1723 he sent written instructions on the czar's orders to an associate, Ivan Tolstukhov: "If mammoth horns are found, an extremely careful search should be carried out so the bones, down to the last of the animal's members, can be collected in their entirety."[23]

In 1722 Messerschmidt sent two "teeth of the mammoth animal" from Siberia to his friend, the German anatomist Johann Breyne in Danzig, along with a brief note in Latin. One was a "molar tooth, apparently diluvial, of an animal that is currently unknown, the mammoth of the Russians—unless it be that of an elephant, a point on which I would like your opinion. It was discovered on a mountaintop near the Thomas River. The other is a piece of ivory tooth that much resembles an elephant's external tooth, which was found among others in the Thomas mountains."[24]

After carefully examining these remains, Breyne presented his thoughts on the "mammoth teeth and bones found underground" to the Scientific Society of Danzig in 1728. The first object studied was a molar "a foot broad, half a foot long, and three inches thick, weighing 8 pounds 3 ounces, pretty entire, except that it is broken in two pieces, and the extremities of the roots are spoiled." The other is "a piece of a *dens exertus* 8 inches long and 3 inches thick, of 1 pound and 6 ounces weight; in some places not different from ivory, but in others calcinated like the common *unicornu fossile*."[25]

From his examination of the remains that Messerschmidt sent him, Breyne concluded that these were elephant bones. "These teeth are mainly found in northern Siberia near such rivers as the Yenisey, the Trugau, and the Lena along the Arctic Ocean, at the time when broken ice erodes the riverbanks and undercuts nearby mountainsides. They are found in such quantity as to satisfy the requirements of commerce and the czar's monop-oly." Sometimes almost complete skeletons were found, and these teeth and bones were not always the same size. Some were very large, and some of the *dentes molares*, or grinders, weighed twenty to twenty-four pounds, ac-cording to Captain Muller. As for the *dentes exerti*, or external teeth, Ides claimed that two of them weighted four hundred pounds, but that others were smaller. Finally, the material of these "exterior teeth" is used to make various objects, such as boxes and combs. Added Breyne: "It is identical to ivory, but somewhat more brittle, and it yellows readily when weathered or heated."[26]

From these remarks, and from having directly examined the remains, Breyne reached two main conclusions:

> I. That those mammoths' teeth and bones are truly natural teeth and bones, belonging heretofore to very large living animals; because they have not only the external figures and proportions, but also the internal structure analogous to natural teeth and bones of animals.
>
> II. That those large animals have been elephants, which appears by the figure, structure, and bigness of the teeth, which do accurately agree with the grinders and tusks of elephants.[27]

Breyne's conclusions matched those that Hans Sloane, president of the Royal Society in London, had presented the previous year to that prestigious assembly and to the Academy of Sciences in Paris.[28]

Back from his expedition to Siberia, Messerschmidt returned to Danzig in 1730 with a series of drawings showing various parts of the skeleton: a very large skull, external teeth, a molar, and a femur "belonging to the animal commonly called mammoth, found in Siberia." The illustrations show a frontal view and two side views (right and left) of the skull, a molar, a tusk, and a femur. Captions in Latin give the dimensions of these remains and describe their morphology in detail. The presence of a molar still well seated in an upper jaw socket served to prove, in a fairly convincing way, that "this skeleton is that of an elephant, and not of the rabbis' chimerical Behemoth." Messerschmidt's drawings were also accompanied by a sworn statement made by one Michael Wolochowicz, who reported the discovery of a piece of skin covered with hairs:

> That head was found by a certain Russian soldier, Basil Erlov, on the eastern bank of the Indigirka River, not far from the mouth of the Volokovoi-Ruczwi stream. After it was discovered, being at leisure, I was present and was an eyewitness to the digging up of this skeleton or bones. And further likewise, on the other bank of the same river, which bank is named Sztanoijahr, I saw a piece of skin putrefied, appearing out of the side of a sand-hill, which was pretty large, very thick, and covered with long hair, fairly thick-set and brown, somewhat resembling goat hair: which skin I could not take for that of a goat, but of the Behemoth, inasmuch as I could not ascribe it to any animal that I knew. . . .
>
> Dated at Irkutsk, 10 February 1724. Michael Wolochowicz.[29]

In 1738 Breyne—by then a corresponding member of the Royal Society in London—sent Sloane a collection of documents relating to the question of the Siberian mammoth: "Mammoth's teeth and bones, a very remarkable and particular curiosity of Russia." These collected papers were published in 1741 in *Philosophical Transactions*[30] and represent the sum of knowledge accumulated in Russia over the previous several decades. For the first time, it presented firsthand evidence of various sorts supporting the thesis that the mammoth remains were those of elephants. And it appeared as the official notice of identification of the "Siberian mammoth," at least until it was re-examined by Cuvier some sixty years later.

Breyne's paper served to identify not only the Russian mammoth, whose ivory the Siberians carved, but also the fragments of "fossil ivory" found buried in the countries of western Europe.[31] The generally accepted explanation was that these remains had been carried and deposited there by the Flood.

> Those teeth and bones of elephants were brought thither by no other means than those of a Deluge, by waves and winds, and left behind after the waters returned to their reservoirs, and were buried in the earth, even near to the tops of high mountains. And because we know nothing of any extraordinary deluge in those countries, but of Noah's universal flood, which we find described by Moses; I think it more than probable, that we ought to refer this strange phenomenon to the said Deluge. In such manner, not only Holy Scripture may serve to prove natural history, but the truth of Scripture, which says that Noah's flood was universal, a thing which is doubted by many, may be proved again by natural history.[32]

Thus, in the first decades of the eighteenth century, the presence of elephant remains in countries as far distant as Ireland, England, Italy, German, Poland, and Siberia became a "natural" proof of the universality of the biblical Flood.

Ever since the Middle Ages, the Flood had been invoked to explain, in particular, the presence of fossil shells on mountains. This explanation was extremely popular during the last decades of the seventeenth century, first in England, then in France and Switzerland. Animal and plant fossils found underground were described as "witnesses to the Deluge" and confirmation of the biblical account. In 1668 the Englishman Robert Hooke first spoke of fossils as "medals of the Deluge."

Inspired by the Cartesian view of the formation of the earth,[33] the English

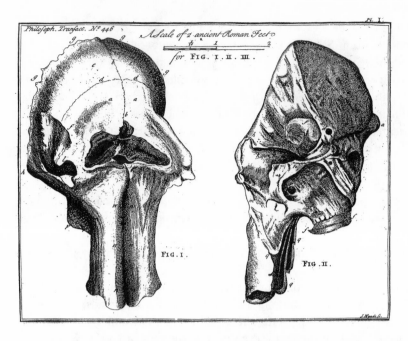

The Russian mammoth: skull, tusk, and femur. The publication of these drawings by the German naturalist Daniel Messerschmidt in the Royal Society of London's *Philosophical Transactions* in 1741 can be considered the "official" eighteenth-century document identifying the mammoth as an elephant. The picture of the skull inspired Cuvier in his first comparative study of mammoths and living elephants in 1796 (see page 109).

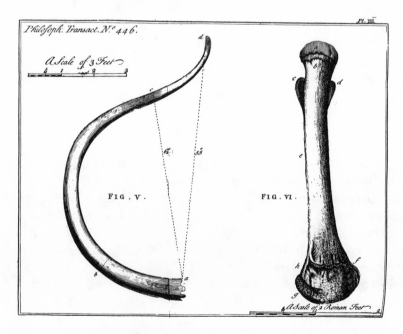

"diluvialists" tried to give reality to Descartes's theory by matching its various stages with the episodes of the biblical narrative. The most famous of these diluvialists was the Anglican minister Thomas Burnet (ca. 1635–1715). In his *Sacred Theory of the Earth*,[34] Burnet wrote that the earth was originally a fluid mass, a liquid mixture of all the elements: primordial chaos. The elements separated out under the effect of gravity, with the heavier ones accumulating in the center to form a hard nucleus, covered by a layer of water, which was covered in turn by a sheet of oil. Particles in suspension in the air, as they fell to the oily sheet, formed a smooth, even alluvial crust on top of the water: this was the land of Eden, a uniform globe without mountains. Heated by the sun, the earth's crust ruptured, and waters from the "great abyss" emerged and mixed with torrents of rain. This "cataclysm"—the only real "event" in this history of the earth—is identified as the universal cataclysm of the biblical Flood and is seen as both physical phenomenon and divine punishment. When the waters retreated "into their caverns and their reservoirs," they revealed the ruins of the collapsed crust, which constitutes the relief of our earth today. Since then the instability of that shattered crust caused the collapses and earthquakes of historic times. To Burnet, the earth today therefore appeared as a heap of chaotic ruins. His thinking about the current appearance of the earth came from his viewing its jagged mountains, and in particular the dramatic relief of the Apennines during a trip to Italy.

Burnet's story, in which he attempted to combine a rational (Cartesian) approach with an apologetic intention and empirical observation, was immensely successful among the naturalists of his day. Other "theories of the earth" were developed along the same lines during the years early in the century. Reinterpreting the Genesis story in rationalist terms, they turned the biblical Flood into the major event of this history. The "diluvialists" in Burnet's lineage were mostly Protestant and English, like John Woodward and William Whiston, or Swiss, like Johann Jakob Scheuchzer, Louis Bourguet, and, later, Elie Bertrand and Jean-André Deluc. They explained the "ruiniform" appearance of the earth as the result of the diluvial catastrophe, and the presence of fossils as "relics of the Deluge."

Physica sacra,[35] by the Zurich doctor Johann Jakob Scheuchzer, is a particularly well-developed example of this approach. Scheuchzer was above all a collector; he had gathered remarkably preserved fossil remains in the rich Tertiary fossil-bearing layers of Oeningen, and his collection and cabinet were famous throughout Europe. But Scheuchzer also considered himself a "thinker," a theologian and man of science versed in natural history and Newtonian physics. He felt the need to use natural history to *prove* the truth of the biblical narrative and to reconcile physics with theology. His goal was to turn Scripture into a commentary that was scientific as well as philological or theological. Inspired by both "the objects of Holy Writ and of nature," Scheuchzer proposed "to shine another feeble ray of light on a few aspects of our sacred writings" by means of the new illumination furnished by the natural sciences. Not that it was necessary to account for divine miracles scientifically: the Creation of the world and the Flood were unique, miraculous events that depended on God's will. But ordinary natural phenomena could be explained rationally. In opposing the separation between scientific knowledge and religious exegesis, which some naturalists had done in the early eighteenth century, Scheuchzer affirmed the compatibility of scientific and religious knowledge. "I have long been persuaded, and have so stated in other works I have given to the public, that one should not view discoveries in physics, mathematics, or medicine simply as curious or useful for the needs of life; they should be applied to practice and to faith; the ideas they furnish must be sanctified and serve us as nourishment not only for the spirit, but also for the will and the heart."[36]

Illuminating the Bible through science: that is the point of this "sacred physics." "Revelation does not exclude reason," wrote Scheuchzer, even if— as he claimed in quoting Bacon—it is sometimes necessary to stray from "the sentiment of the church fathers" regarding things that belong to the domain of reason. For those with real faith, the scientific attitude is justified in that it allows "the assiduous contemplation of the marvels of creation and of Providence." Like the "theologies of insects" of René-Antoine Réaumur, abbé Pluche, and all those who took the work of naturalists as an occasion to marvel at divine wisdom and intelligence in these first years of the eighteenth century, Scheuchzer proposed a theological approach to natural phenomena, and notably a "theology of fossils."

Fossil objects, he said, were the archaeological witnesses to this history— "relics of the Deluge." Curiosities dug up from the earth, plants, fossil bones, and shells, all these "marvels of Creation" refer directly to the biblical story. "These are the very originals which were buried in the layers of the earth at the time of the Flood."[37] Scheuchzer's cabinet (and especially his famous collection of fossils from Oeningen schists) was a store of "marvels" and curiosities, but it also was a *reliquary*, which could serve to support a demonstration of a truth both scientific and theological. For these "monuments"—the fossil remains—can properly be called "relics," witnesses of ancient history as told in Holy Scripture; they have the sacred character of objects of those ancient times described in the Bible.

Only the diluvial explanation could account for the presence of animal remains, of "entire bones found in the bowels of the earth and sometimes embedded in the very rocks": terrestrial animals, but also marine ones. "For how could all these animals, especially the fishes of the sea, be found in places so far from the sea, and even inside very high mountains, like those in Switzerland, if not through the universal Flood?"[38]

In presenting these "relics" of the Deluge, Scheuchzer enumerated the fossilized objects in his cabinet of curiosities, carefully organized by class. His text is primarily a list of objects, a stroll among the numbered specimens, which are described and presented in a short introduction that relates them to the work's central "philosophical" (or theologico-scientific) thesis. A visit to Scheuchzer's collection began with those objects "that belong to the vegetable kingdom." "Here one can see wood, leaves, fruits, and entire plants imprinted in the stones, and found in the different layers one encounters when digging in the earth."

In his *Herbarium diluvianum*,[39] Scheuchzer presented animal remains and fossilized plants, in particular an ear of barley whose degree of maturity suggests that the Flood took place in the month of May. The largest part of this collection consisted of the remains of marine animals, fishes, and shells. One could see the remains of a crocodile and a viper, and even "a bird's tail feather . . . which is perfectly imprinted on slate from the Oeningen quarries." In underlining this particular vestige's rarity, Scheuchzer returned to the diluvial explanation: given the "lightness of birds," he explained, one understands that "little or almost nothing of the monuments of the Flood remains." In the same way, fossilized insects are very rare in the cabinets of the curious, but Scheuchzer possessed an "escabot in an Oeningen slate," "a dragonfly from the Verona region," "an elephant molar, or large tooth," and "the teeth of several other quadrupeds found in the earth." Finally, Scheuchzer had one object rarer and more precious than all the rest: the skeleton of the famous "witness to the Deluge," one of the people who were "the main cause of the worldwide destruction, and for whose sins so many other innocent victims had to be sacrificed." This was *Homo diluvii testis*, which Cuvier would identify a century later as the remains of a giant salamander.

In his approach as both a naturalist and an apologist, Scheuchzer precisely named fossil species and identified them as analogs of existing species. As for those that did not seem to have a modern counterpart, they were believed to exist deep in the sea, hidden to our eyes, but certainly present: "Ammon's horns [ammonites] are unlike any shells found in the [natural] history of the sea, but doubtless the same species, in the same variety that we dig from the earth, are to be found at the bottom of the sea; except that those on the sea floor cannot be washed ashore, even during the fiercest storms."[40]

To Scheuchzer's eyes, no animal in the present world could exist that did not exist in the first days of Creation. Fossils of "exotic" species found in our regions could have been carried by the waters of the Flood when they covered the entire earth. Henceforth, the role of the diluvial explanation was not only

The earth after the Flood, according to Scheuchzer, *Physica sacra*, vol. 1 (1732). The tortured relief of today's mountains and the presence of fossils within them were interpreted as resulting from the diluvial cataclysm. (Bibliothèque Nationale de France photo, Paris.)

to account for the existence of fossils within the earth, but also to explain their being borne to very distant places.

By the mid-eighteenth century in western Europe, fossil ivory had evolved from being a pharmaceutical product to a material prized in art and craft, made into combs, "handles for sabers and knives, boxes,"[41] and so on. In 1762, in his *Dictionnaire universel des fossiles*,[42] the Swiss pastor Elie Bertrand summarized what was known about "fossil ivory" at the time. He still used the term *unicornu fossile* but preferred the Russian *momontavakost*, which by then seemed to have been almost universally adopted. This referred to mammoth bones, most often found in Siberia, "especially along several rivers, the Lena, the Yenisey and the Obi," but that have also been found in Switzerland "in the canton of Bâle." According to Bertrand, this fossil ivory, like natural ivory, came from elephants, from their "molars and incisors," to be precise, and very probably from males, since the females do not have tusks. These teeth have even been seen "still attached to the sockets of a jawbone." Referring to the many people who traded in this precious material, he noted "the differences that one observes between fossil ivory and natural ivory":

1. Fossil ivory is covered with a shell of yellow-gray, whitish, or greenish crust.
2. It is white on the inside, but marked with black dots.
3. It smells like almond milk.
4. It tastes like white chalk.
5. It is as hard within as it is without.
6. It is easily cut into sheets or slices.
7. When one dips it in water, it causes the water to foam. It sticks to the tongue like marl or wood.[43]

In general, these remains could be compared to the tusks of elephants living in India or Africa. African elephants, "especially those from Bombaze [Mombasa] and Mozambique," probably produced the fossil tusks found in Siberia, because Africa has "such a quantity of elephants that one sees wandering in huge herds." To explain the transport of these herds of elephants from Africa to Siberia, the diluvial thesis is de rigueur. "The waters of the Flood could have carried these prodigious herds to those countries where their teeth have been found."

By the second half of the eighteenth century, "fossil ivory" had lost almost all its mystery. Anatomical comparison and the careful examination of remains furnished the elements of this identification. All that was left was to explain the presence of these warm-climate animals' remains in the frozen soil of Siberia. And here the diluvial explanation came readily to Bertrand, as it had to the travelers in Siberia—Ides, Tatischev, Gmelin, Pallas, and Messerschmidt.

Yet this was already the subject of heated debate.

In d'Alembert and Diderot's *Encyclopédie*, the Flood was not used to explain the presence of fossils, shells, or the bones of land animals found in the layers of the earth. For the previous several decades, the diluvial "systems" hatched in the late seventeenth century in England had been facing new questions born of more careful observation and a rational critique of religion as a whole. The "Deluge" entry, mainly written by Nicolas-Antoine Boulanger and revised by Baron d'Holbach, recognized that a flood might have been responsible for some aspects of the earth today, but not for the presence of fossils. "The earthquake that shattered Mount Ararat, and left it with such a hideous and frightening aspect, was not capable of inserting fossils into the broken remains. The agency that separated Europe from Asia at the Bosporus did not bury those marine remains found in the country's interior into the cliffs and escarpments left on either side of the isthmus."

As an account for the processes of the history of the earth, the biblical Flood's influence was clearly beginning to ebb. If conceived as divine punishment, it had to be represented as a cataclysm, and could therefore not be responsible for the nature, presence, and disposition of fossils preserved deep in the earth. This was especially true of fossil shells, some of which were completely unknown or belonged to species that might exist only in distant seas. Moreover, their abundance and distribution could not be explained by the diluvial thesis of a brutal upheaval. This critique of an irrational and miraculous event—the Flood—as a major event in the earth's history also undermined the hypotheses about the Siberian mammoth.

The *Encyclopédie* defines "fossil ivory" as "enormous teeth resembling large horns, which have often been found underground." The variations in their appearance (they are white, yellowish, or brown; sometimes very hard, at other times more fragile and breakable) are due to "the greater or lesser degree of decomposition that these teeth have suffered in the various places in the earth where they were buried."[44] They are found in England, Germany, France, and even in the Grenelle Plain, "that is, at the gates of Paris." But it is especially "in Russia and in Siberia, where these bones . . . have been uncovered by the waters of the great Lena and Yenisey Rivers . . . which erode the earth from the banks when they carry large pieces of ice during the spring breakup."

How does one account for the immense quantity of fossil ivory found in regions so far to the north? One hypothesis was that they were elephants brought from Africa by the Roman armies. But while that might account for elephants found in the soil of Italy or Europe, "these conquerors never made war against the hyperborean Scythians and it does not seem that any other Indian conqueror was tempted to wage war in such a harsh climate so far away. . . ."

The explanation for the elephant remains in Siberia had to be found elsewhere. The author of the article reached a twin conclusion: a change of climate may have been involved ("one must therefore conclude that in a period now lost to history Siberia enjoyed more benign weather") or else "a general revolution of our globe," which "buried [these animals] in the bowels of the earth." Between these two explanations, there was no room for the diluvial hypothesis (beyond its verbal echo of a "revolution").

By midcentury the fantastic edifices of the diluvialists had been rejected, replaced by the requirements of a more rigorous description of natural phenomena. Henceforth, it was possible to debate the formation of the globe as an independent matter. The separation of faith and reason allowed people to think about the laws that govern the history of the earth and the universe without bringing in irrational or miraculous causes.

In his 1749 *Theory of the Earth*, Georges-Louis Leclerc de Buffon launched a vehement criticism of the diluvial "systems," refusing to allow science to be turned into a "novel." After setting out the diluvial theses of Burnet, Woodward, and Whiston, he severely judged them as "made-up and fabulous stories" by turn-of-the-century naturalists trying to match their views to the letter of the Scriptures. "Whenever one is bold enough to try to explain theological truths by physical reasons, or to interpret the divine texts of the holy books from a purely human viewpoint and reason about the will of the Almighty and the execution of his decrees, one necessarily stumbles into darkness and chaos."[45]

Still, the diluvialist explanation did not disappear with Buffon's criticism. The ambiguous notions of the *diluvium* and of "antediluvian" animals would persist well into the nineteenth century, in the works of Cuvier, William Buckland, and Boucher de Perthes. In his 1887 book *The Mammoth and the Flood*, the English naturalist Henry Howorth developed at length the idea that the mammoths of Siberia drowned in the biblical Flood. His suggestion that the name "mammoth" had a biblical origin (Behemoth) reflects the tradition, which would persist until the end of the nineteenth century, of trying to fit paleontological and prehistoric knowledge into the narrative framework of Holy Writ.

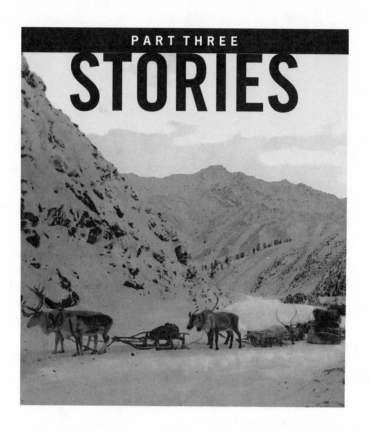

PART THREE

STORIES

The *"Vast* Mahmout" and the Birth of the American Nation

When Thomas Jefferson was the governor of Virginia, he listed the animal and vegetable species common to America and Europe in his 1781 book, *Notes on the State of Virginia*.[1] In the list of New World animals, he insisted on including the mammoth. Jefferson—one of the actors of the American Revolution, a drafter of the Declaration of Independence, and a future president of the United States—firmly believed that mammoths were living somewhere on American soil. To skeptics who asked him why, he is said to have calmly answered, "Why not?" The motives behind Jefferson's belief were deep-seated ones, in which scientific reasons joined politics and national pride.

Is it too much to say that the image of the mammoth is closely associated with the heroic period of the building of the American nation? During the last half of the eighteenth century, the lively debate over the question of this animal's existence on American soil involved not only the history of the exploration and conquest of vast, as yet unknown territories and epic battles between colonists, but also the peaceful or bloody relationships among various groups of whites and Indians, and political and scientific exchanges between America, France, and England. Adding to the debate immediately following the American Revolution were the elements of nation building: an awakening of national feeling and the birth of scientific and scholarly institutions.

Notes on the State of Virginia is an essential document on the political and scientific history of the United States in the first years of its existence. Begun in summer or fall 1780, the book was the result of an inquiry made by the French authorities that had supported the American Revolution. François de Barbé-Marbois, the secretary of the French delegation to Philadelphia, gave Jefferson a list of twenty-two questions related to the political organization, geography, and fauna and flora of each of the new republic's states. Jefferson undertook to gather firsthand information on these various subjects and to organize them in the form of a book, adding his own observations and commentary. The interest he brought to this task was no doubt due to his responsibilities as governor of Virginia, but also to the fact that in early 1780 he had been elected to membership in the oldest and most prestigious American society, the Philosophical Society in Philadelphia (he would become its president the following year). It is precisely in the context of this inquiry that, starting in December 1781, Jefferson became fascinated by "the gigantic animal whose remains have been found on the Ohio River." The affair was already several decades old, since, as we will see, the first remains of this mysterious animal had been collected as early as 1739.

Before 1781 Jefferson had never taken any special interest in fossil bones. But from then on he became passionate on the subject, maintaining a correspondence with the explorer Colonel George Rogers Clark, who sent him many specimens, including "a femur, a jaw with molars, and a tusk." Because of his determination to discover the truth about the American mammoth, Jefferson played an important role in the early days of American paleontology. For thirty years he would encourage and personally finance fossil research on the American continent, collect Indian stories and legends, maintain a voluminous correspondence on the subject, help expeditions financially, promote interest in the study of fossils (notably within the Philosophical Society), and contribute to the conservation of remains.

In them, Jefferson saw the mortal remains of an animal he thought might still be roaming the empty, unexplored parts of the American territory. To him, this was the American brother of the mammoth, that gigantic animal that had never been seen alive but whose remains had been noted by every European voyager to Siberia for the previous century. Jefferson's belief rested on several bases. First, for philosophical and religious reasons, he refused to believe in the extinction of species. "No instance can be produced, of [nature] having permitted any one race of her animals to become extinct; of her having formed any link in her great work so weak as to be broken."[2]

The argument is a familiar one: If God—or "nature"—has created the order of the world and fixed species in the hierarchy of beings for all eternity, one cannot deny the power to maintain this order as it was originally established. To think that species could be "lost" or become extinct would be to deny the perfection of the world and the sacred links that join beings to one another. The fact that these animals had not yet been seen did not prove their disappearance, only our ignorance. A good lawyer, Jefferson refused to accept negative arguments as proof. "The traditional testimony of the Indians, that this animal still exists in the northern and western parts of America, would be adding the light of a taper to that of the meridian sun. Those parts still remain in their aboriginal state, unexplored and undisturbed by us, or by others for us. [The mammoth] may as well exist there now, as he did formerly where we find his bones."[3]

In his quest for evidence of the American mammoth, Jefferson collected Indian legends, in a true work of ethnographic inquiry. Certain legends spoke of an animal that the Indians called the "Great Buffalo," and whose bones and teeth they knew well. These were found in abundance along the Ohio River, in particular at a now-famous site, a salt marsh where great mammals came to lick the salt-impregnated soil. This location, which is well-known because of the great number of large bones found there, sometimes right on the ground, is called Big Bone Lick. While Jefferson was governor of Virginia, a visiting Indian delegation told him of a story shared among the tribes:

> That in ancient times a herd of these tremendous animals came
> to the Big-bone licks, and began an universal destruction of the
> bear, deer, elks, buffaloes, and other animals, which had been
> created for the use of the Indians: that the Great Man above,
> looking down and seeing this, was so enraged that he seized his
> lightning, descended on the earth . . . and hurled his bolts
> among them till the whole were slaughtered, except the big
> bull. . . . Springing round, he bounded over the Ohio, over the
> Wabash, the Illinois, and finally over the great lakes, where he
> is living at this day.[4]

Intriguing though these accounts were, they remained tenuous. But Jefferson's eagerness to prove the current existence of this fabulous animal was spurred by other serious reasons. In affirming the mammoth's existence among American fauna, Jefferson was responding to a claim by the French naturalist Buffon that had caused a great deal of talk on both sides of the Atlantic and had wounded his patriotic pride.

In the ninth volume of Buffon's *Histoire naturelle*, which was published in 1761, he described the geographical distribution of the world's animals and compared the Old and New World species. He concluded that the New World animals were smaller and more feeble that those of the Old. As for men, their fate was no more enviable: "There is something in the New World that is contrary to Nature's growth," he wrote.

> There are obstacles to the development and perhaps even the
> formation of great seeds. Even those which got their start or
> grew to their full size under the gentle influences of another cli-
> mate become smaller and shrunken between that thin sky and
> empty earth, over which wander only a few, scattered men. Far

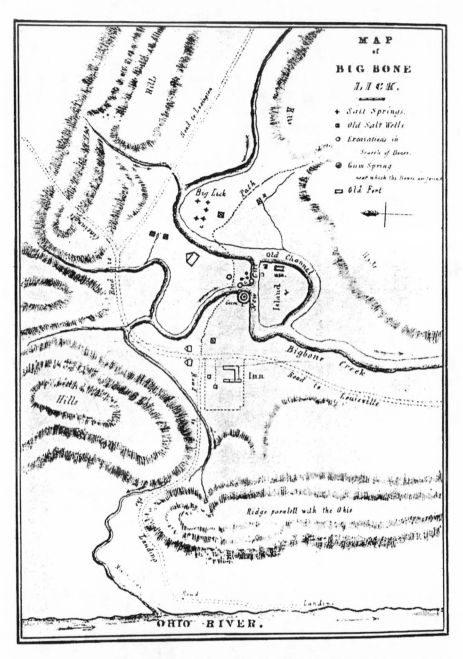

Map of Big Bone Lick, an Ohio River Valley fossil site that became known in the first third of the eighteenth century. This 1831 map shows the site as it appeared in 1828. (Simpson, 1942).

from ruling this territory as his domain, he held no sway over it. Having neither conquered the animals or the elements, mastered the sea, controlled the rivers, or worked the ground, he was but an animal of the first rank, and existed for Nature only as a being of no importance, a kind of impotent automaton, unable to either change or help her.[5]

Buffon went on, in Jefferson's translation:

Although the savage of the New World is about the same height as man in our world, this does not suffice for him to constitute an exception to the general fact that all living nature has become smaller in that continent. The savage is feeble, and has small organs of generation; he has neither hair nor beard, and no ardor whatever for his female.[6]

In the mid-eighteenth century, Buffon was France's greatest naturalist. He had been a member of the Académie des Sciences since 1733, and since 1739, the director of the Jardin du Roi, a world intellectual center for the natural sciences. Respected by the French court and European sovereigns, Buffon wielded great influence in France and Europe. His monumental *Histoire naturelle*—whose volumes, illustrated by magnificent engravings, were published starting in 1749[7]—had given him immense notoriety.

Buffon's statements about the feeble character of American species soon became known beyond the Channel and the Atlantic. In his unflattering picture, one can recognize an application of Buffon's philosophy of nature: Living species are subject to the influence of climate, of "circumstances," and of living conditions. Changes in nature have an influence that can cause living beings to "degenerate." Buffon had set forth his ideas on the "degeneration of animals" at length in 1766. America's climate was rougher, colder, and more humid than Europe's, he claimed, and particularly unfavorable to living beings. Human action on nature can modify the conditions of the milieu, but the American territory was wild and the animals feeble because men, who had scarcely risen above their original condition, were scattered on a hostile land and were too uncivilized to improve their own lot, much less that of animals. Today's American species, assuming they had the same origins as European ones and had later been separated by "immense seas and trackless wastes," had been "shrunken and denatured" by the influence of a climate that was too harsh, untempered by human action. To Buffon, only Europe and its moderate climate were worthy of being the home of well-proportioned and vigorous beings. America's natives, and even the colonists who settled there, could only be—or become—the representatives of an inferior, degenerate humanity, and the same was true of animals. Buffon was obviously injecting European prejudices into his scientific reasoning—a naturalist's speculation supporting racist and colonialist arguments.

It is therefore easy to understand Jefferson's determination to prove the existence of a living mammoth on American soil: his country and his compatriots' honor were at stake. Turning Buffon's argument around, Jefferson tried to demonstrate—with measurements to prove it—that America's humidity level was no higher than Europe's and that his country's soil bore animals as large as those known in Europe.[8] And he could point to one obvious proof: the discovery of gigantic bones that had been made forty years earlier in the Ohio Valley.

In 1739 Baron Charles de Longueuil, a French-Canadian colonist and a major in the army of Louisiana, was placed in command of French and Indian troops and dispatched to help Jean-Baptiste Le Moyne de Bienville, the founder and governor of New Orleans, in an attack on the Chickasaw Indians. He brought back from the campaign some very curious bones. The expedition, which left Montreal in June 1739 and headed for the Ohio River, followed the river and reached the area near present-day Louisville in late summer. On the edge of a swamp, they found what appeared to be the remains of three elephants. Longueuil had parts of these remains collected, including a tusk, a femur, and at least three molars. These were carried with his army and—after the victory over the Chickasaw—sent to Europe in 1740.

What animal could the remains collected in the Ohio Valley belong to? Longueuil had quickly identified them as those of elephants, but the matter turned out to be far more complicated for the French scientists to whom the young officer and casual naturalist sent his booty.

The bones Longueuil collected and sent to the Cabinet du Roi wound up raising serious questions in the minds of the three most brilliant naturalists of the late eighteenth century: Jean Guettard, a Swiss geologist and mineralogist; Louis Daubenton, a young anatomist and zoologist whose career would outlast the French Revolution; and Buffon, the famous intendant of the Jardin du Roi.

In a paper presented to the Académie des Sciences in 1752[9]—and that included the first geological map of America—Guettard described one of the molars and said exactly where it came from: "a place that is marked on maps of Canada as the canton where elephant bones were found."[10] He gave a picture of the molar but did not try to identify the animal it belonged to. "What animal does it come from?" Guettard wrote. "And does it resemble fossil teeth of that size which have been found in various parts of Europe? Those are two points on which I was unable to shed any light."

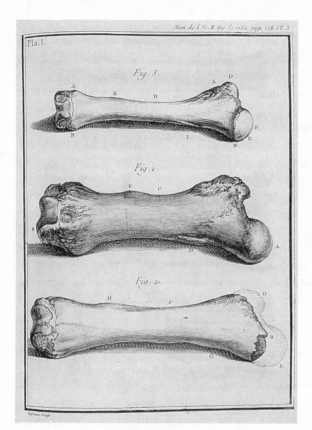

Daubenton's study comparing the femur of an elephant (3) with that of the Siberian mammoth (2) and the "unknown animal of the Ohio" (1). This plate shows one of the first anatomical comparisons between living vertebrate and fossil bones. Unlike Cuvier, Daubenton concluded that the three bones belonged to the same species. (*Mémoires de l'Académie des Sciences* [1762], 228, pl. 13.)

When Daubenton restudied the remains ten years later, in 1762, he started by examining the bones. Looking at the animal's femur, he had the idea of comparing it to the femurs of an elephant and of a Siberian mammoth. They appeared in an engraving that accompanied Daubenton's article, and the comparison between the three bones is particularly eloquent: despite their differences in size, they have the same characteristic shape. The demonstration was brilliant, and the concept—in mid-eighteenth century—quite novel. It involved the same kind of anatomical comparison that would allow Cuvier to identify and reconstruct extinct species a few decades later. But Daubenton did not consider the extinction hypothesis. For him, the three similar femurs belonged to three animals of the same species, three variations on a single type.

The study of the tusks, which were somewhat different in size and shape, confirmed the conclusion that could be drawn from the femurs. Daubenton had noted the material and structure of ivory from living elephants and that found in Siberia. He had studied the different colored "fibers," the web of lines that start from a black dot known as the tusk's "heart" and that form lozenges where they intersect. He concluded that the structure of elephant tusk ivory was exactly analogous to that of the Siberian mammoth.[11] As for the Ohio River animal, its tusks certainly resembled those of known species of elephants. Examining the molars, however, plunged Daubenton into perplexity. The square, bumpy teeth could not be those of an elephant, but on close examination, resembled hippopotamus molars, having the same clover-leaf pattern on wear surfaces. Daubenton concluded that they were the teeth of a gigantic hippopotamus, which had become mingled at the site with the bones of an enormous elephant. The idea that the bones, tusks, and teeth all belonged to a single animal was supported only by the claims of the "ignorant savages" who had discovered the remains. The proper scientific approach was to separate them.

Daubenton's demonstration lacked neither finesse nor perspicacity, but it stopped short of the point where a totally new hypothesis was required—the extinction of certain species. This he found unthinkable. Daubenton was a fixist, and his approach consisted primarily in viewing the remains entrusted to him as those of animals that currently existed. From that perspective, it would be inconceivable or even absurd to imagine an animal whose species had vanished. As George Gaylord Simpson wrote, Daubenton had reached the "limits of scientific imagination" of his day.

In the middle of the eighteenth century, new discoveries had revived the debate over the "unknown animal of the Ohio." The "exploitation" of the famous Big Bone Lick site probably began in the 1750s and would yield a very large number of bones, teeth, and tusks. Several decades later, the animal they came from would become known as the American mastodon.

These discoveries were made against the background of the exploration of the country and struggles between different groups of whites and Indians. Driven from their territory by the French, English colonists and traders settled in the upper Ohio Valley in the 1760s. The region's most famous fossil collector was William Croghan, an Irish Protestant born in Dublin whose family settled in Pennsylvania in 1741,[12] and who had gotten to know the Iroquois and Delaware Indians. In 1765–66, during the conflicts between French and English colonists, he was sent to "treat with the Indians who retained French sympathies and to attempt to open the Illinois region to British penetration."[13] During this expedition, Croghan found "a prodigious number of bones of very large animals, which by the size and shape of the bones and tusks, lead one to conclude that they are the bones of elephants."[14] The site was near "a large salt marsh where wild animals congregate at certain times of year," southeast of the Ohio River, "about 640 miles from Fort Duquesne." After a number of trying adventures, during which he was held captive by Indians, Croghan reached New York. From there, he shipped his paleontological booty to the English political and scientific authorities in 1767. Lord Shelburne, who was in charge of the American colonies, received "two large tusks, one of which was quite whole and was nearly 7 feet long . . . a jawbone with two teeth, and several very large molar teeth,"[15] which he sent to the British Museum.

Most of these remains reached London in 1767, and over the following months they generated lively interest on the part of English naturalists. On February 25, 1768, William Hunter presented his study of the remains of *Ohio incognitum*. From his examination of the tusks, lower jaw, and molars, Hunter concluded that the remains belonged to a single animal of a species different from the modern elephant: in his opinion, a kind of carnivorous elephant. "And if this animal was indeed carnivorous, which I believe cannot be doubted," wrote Hunter, "though we may as philosophers regret it, as men we cannot but thank heaven that the whole generation is probably extinct."[16]

Benjamin Franklin, who was in London at the time, also received part of the booty. On August 15, 1767, he wrote Croghan to thank him for sending him the remains. He set out several hypotheses on the nature of the animals they came from, but also noted several enigmas. Like Hunter, he initially thought that a carnivore was involved. But how could one explain such an accumulation in a single area of bones of such an enormous animal, which had never been seen on the American continent, and of which similar remains were only found in Peru? Some parts of the animal (notably the tusks) were similar to those of elephants, which "now inhabit naturally only hot countries where there is no winter." How could such an animal survive in areas with winter, like the Ohio Valley, not to mention Siberia? It looked, concluded Franklin, "as if the earth had anciently been in another position, and climates differently placed from what they are at present."

Yet in a second letter a few months later, addressed to abbé Chappe d'Auteroche on July 31, 1768, Franklin returned to his first hypothesis about the "Ohio animal." He no longer believed that the molars were those of a carnivore.

> Some of our naturalists . . . contend that these are not the
> grinders of elephants but of some carnivorous animal unknown,
> because such knobs or prominences on the face of the tooth are

not to be found on those of elephants, and only, as they say, on those of carnivorous animals. But it appears to me that animals capable of carrying such large heavy tusks must themselves be large creatures, too bulky to have the activity necessary for pursuing and taking prey, and therefore I am inclined to think those knobs are only a small variety. Animals of the same kind and name often differing more materially, and that those knobs might be useful to grind the small branches of trees, as to chew flesh. However I should be glad to have your opinion, and to know from you whether any of the kind have been found in Siberia.[17]

In changing his mind about the animal's diet, Franklin was following a line of reasoning that applied not only to the tooth, but to the entire organism. The elephants of the Ohio were only a "small variety" of those that exist today, and perhaps of those whose remains are found in Siberia. Their teeth would have been adapted to a particular diet. In this, Franklin was perhaps influenced by another English naturalist, his friend Peter Collinson. Collinson had also studied the Croghan remains and delivered two papers to the Royal Society in London, on November 27 and December 10, 1767. From his study, Collinson reached three main conclusions. First, the remains definitely belonged to a single animal, because teeth like the "bumpy" molars were always found along with tusks and remains that resembled those of an elephant. Second, a careful examination of the anatomy of the animal's teeth revealed that it must have been an herbivore, capable of chewing leaves and branches. Finally, this animal, which was doubtless related to the elephant, was currently unknown.

These conclusions were submitted to Buffon, the era's highest scientific authority in these matters. In a letter he wrote Buffon on July 3, 1767, Collinson presented his conclusions in the form of paradoxes. The tusks may resemble those of elephants in all respects, though often much bigger, but the molars don't resemble elephant teeth at all. "How is one to reconcile this paradox?" asked Collinson. "Could one not suppose that there once lived a large animal that had elephant tusks and hippopotamus grinders?" This was a way of making explicit the hypothesis of the existence of animals whose "species was lost," and even hybrid, monstrous animals whose appearance would be very different from that of any known animal today. Moreover, there were no elephants in America; they lived in hot climates and were found in Asia and Africa. How could one explain finding remains that were clearly related to elephants in countries where winters are very harsh?

Collinson, Franklin, and Hunter had reached the same conclusions, that the remains were those of animals that belonged to an unknown, extinct species, which had probably gone extinct because of a change in climate. Buffon's speculations would turn out to be quite different.

Buffon answered Collinson's questions, which he quoted verbatim in footnotes to his *Epochs of Nature*, in three ways. First, he wrote, the teeth and tusks do not belong to the same animal.[18] The bones found at Big Bone Lick were the mixed bones of elephants and giant hippopotamuses. The elephant can be recognized by its tusks and—as Daubenton stated—by the fact that "the large square teeth whose grinding surface is cloverleaf-shaped have all the characteristics of hippopotamus molars."[19] Among the remains, one must also note the presence of a third animal, as yet unknown, with "enormous teeth, whose grinding surface consists of large, worn points." The drawing that appears with footnote 7 in *Epochs of Nature* clearly shows the posterior molars of the same animal—which has since been identified as a mastodon. For Buffon, however, a "lost species" was obviously involved: "I think I can state with confidence that this species of very large animal is lost," he wrote.[20]

In short, Buffon found three different animals where others saw only one! He refused to accept as proof the fact that these remains were always found

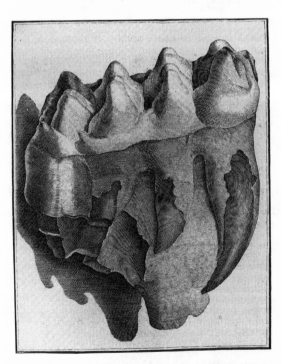

Bumpy molar of the "unknown animal of the Ohio" studied by Buffon, published in his *Epoques de la Nature* (1778).

together, which many American and English searchers and naturalists had noted. "Wherever these grinders are found, there also we find the tusks and the skeleton," wrote Jefferson. "It will not be said that the hippopotamus and the elephant came always to the same place, the former to deposit his grinders, and the latter his tusks and skeleton. . . . We must agree then that these remains . . . are of one and the same animal."[21]

Which left Collinson's second paradox: How to explain the presence of elephants (since they were elephants, and even hippopotamuses) in these wintry countries? To this Buffon added another question, which Collinson had implied: How could one explain the disappearance of certain species? This double inquiry called for more than a quick, easy answer. "What system can be devised, and with what degree of probability that would account for the presence of bones in Siberia and in America," wrote Buffon, citing Collinson's "second little memoir."[22] The "world system" and the history of the earth presented in *Epochs of Nature* in 1778 were Buffon's answer to that question. It is not too much to say that the question of the Siberian mammoth and the "unknown Ohio animal" is the keystone—maybe the key—of this masterwork, which Buffon published at the age of seventy-seven, and which is unanimously considered to be his scientific legacy.[23]

Buffon answered Collinson's queries by telling an immense, impressive story: that of the history of the earth and its creatures.

Nature is not immutable, according to Buffon. It is transformed over the course of time along an irreversible history. Just as historians define "epochs" of human history, nature's "epochs" are characterized by successive phases of its transformation. So it is possible to write an account that describes this history. In the *Epochs of Nature* account, elephants have a central, vital place. They are discussed at length in the "first discourse" and the "fifth epoch" and are the subject of lengthy commentaries and documents, which Buffon added to the footnotes that accompany those chapters.[24]

To sum up, elephants are notable in nature today by their large size and their presence in hot countries close to the equator. But their fossil remains (and it is now essential for Buffon to combine the Siberian "mammoths" and the remains found in Europe or North America in a single species, similar to living elephants) prove that these animals once lived in the northernmost regions of the world. In fact, everything points to a permanent, prolonged stay there, as proven by the enormous quantities of ivory from those parts, which had already been traded in Europe for nearly a century. "More ivory has probably already been brought out of the north than could be supplied by all the elephants living in India today," wrote Buffon. Therefore, the problem must be boldly recast, rejecting Pallas's and Gmelin's diluvial explanation[25] and dealing with the issue of fossil elephants in Siberia in terms of these animals' characteristic adaptation to a given climate. "The question . . . consists in determining if there is or has been a cause that could have changed the temperature in different parts of the world, to the point where northern lands, which are very cold today, could have once been as warm as countries in the south."[26]

The idea was simple and seemed brilliant: The animals didn't migrate northward; the climate changed. But what would explain the change of climate? There were two possible answers. On the one hand, there was the hypothesis of a "change of the angle of the ecliptic,"[27] and therefore the position of the poles, which was put forward in 1714 by Chevalier de Louville. But Buffon refused to take that hypothesis seriously, because the variations of the ecliptic angle occur within very narrow limits. A simpler hypothesis was that the earth had slowly cooled from its initial state as an incandescent globe to its present climate, and that was the one Buffon chose. It "condensed" and reorganized two ideas that he had long been working with: the thesis of an earth in fusion, set forth in *Théorie de la terre*, which was written in 1744 and completed in 1749,[28] and the effect of climate changes on the transformation of living creatures, which Buffon first presented in his *Histoire naturelle de l'homme* in 1749.

To Buffon's mind, a white-hot earth had slowly cooled and was still cooling. Northern countries were the first to become habitable; being less exposed to the rays of the sun, they cooled faster, so they were populated by animals adapted to hot climates. Elephants lived in the countries of the north at a time when they were hotter. And since we also find their remains "in Poland, Germany, France, and Italy,"[29] we can imagine that as the northern regions gradually cooled, these animals migrated "toward temperate-zone countries where the heat of the sun and the globe's greater thickness compensated for the earth's loss of heat."[30]

In fact, the title of the fifth "episode" in *Epochs of Nature* is "When Elephants and Other Southern Animals Lived in Northern Lands." Elephants migrated from north to south as the climate became colder. This "biogeographical" thesis explained the presence of fossil elephant remains not only in Europe's soil, but also in North America, to which European elephants had been able to migrate before an ocean separated the two continents. They were unable to reach South America, Buffon claimed, because they were stopped by the "very high mountains" near the isthmus of Panama. "Elephants were not able to overcome this impassable barrier because of the great cold that is felt at high altitude."[31]

There is another essential characteristic of these animals that lived during earth's burning past, which we have already noted: their great size. "Monstrous," "prodigious," "colossal" are adjectives that Buffon used to characterize these animals from ancient times. And he set down the following law:

"Everything colossal has been formed in the north."[32] Buffon queried ivory artisans and merchants in order to compare the length and thickness of fossil tusks from Siberian mammoths—which they called "baked ivory"—with "raw ivory" from today's elephants. Everything suggested that the ancient animals were much larger than today's species, even though they were much alike. Their tusks measured eight to ten feet long, and the femurs were "at least as long as a modern elephant's, and much thicker." From the beginning, the type endured, but its size changed. Buffon thus removed the traditional theme of giants as the world's first inhabitants from the realm of religion. Perhaps inspired by Lucretius, he wrote: "Nature was in her first vigor; the internal heat of the earth gave her productions all the energy and the range of which they were capable."[33]

Buffon believed in the possibility of species disappearing.[34] As early as his 1749 *Théorie de la terre*, he stated that many marine species had been lost, among them the famous ammonites, the belemnites, "the ortocératites, the lenticular and numismatic stones." The very fact that it was possible to name several lost species proved that marine animals had appeared long before land animals. Among the latter, only a single species had truly disappeared: "the largest of them of all, larger even than the elephant,"[35] and therefore obviously the oldest, the one whose "nature required greater heat than is current in our torrid zones." This was the famous "unknown Ohio animal," with its "huge, square molar teeth with their large worn points," which perished as a victim of the climate's cooling.

As a disciple of Newton, Buffon aspired to explain natural phenomena by physical laws. For him, the causes that rule the history of nature could not be accidental, singular events. They must be ongoing causes, including those at work in the past. "One can . . . compare nature with herself and go back from her current, known state, to several epochs of a more ancient state." In his "actualism," Buffon refused to accept miracles or divine intervention as part of scientific explanation.

In *Epochs of Nature*, Buffon also gave the earth's history an immense time span. The result of experiments and calculations, this further distanced him from religious dogma. During the summer of 1767—right after receiving Collinson's letter with the "paradoxes" about the Ohio animal—Buffon undertook his famous experiments on the heating[36] and cooling of metal in the steelworks on his estate at Montbard, near Dijon. In order to calculate how long it would take the earth to cool, he had iron balls and pieces of metal heated white-hot and noted how soon they could be safely touched and how long it took them to cool completely. He also measured how long it took molten iron to cool.[37] From these calculations, Buffon published his estimate of the earth's age at seventy-five thousand years in the printed editions of

Epochs of Nature. But in his manuscripts,[38] he hypothesized a period forty times longer: 3 million years. Whatever his reasons for making the change (caution, or plausibility?), his estimate of the length of the earth's history was very different from the "six or eight thousand years" suggested by biblical chronologies.

The theme of the cooling of the world underlay a history, the most grandiose history possible, since it unfolded in seven stages—"epochs" in Buffon's terms—from the origins of the earth to the advent of man: an account that has a direction, and whose successive episodes he related. Originally, the earth consisted of an incandescent mass, which then cooled. Its elements hardened and became "vitriscible." An initial general flood occurred, as shown by fossil shells, and when the waters retreated, they left organic remains in the sediments. The next stage was marked by the appearance of great quadrupeds; the distribution of their fossil remains throughout the Old and New Worlds proved that at one time the continents were not separated by the sea. The sixth epoch, the penultimate of this progression, was the one in which "the separation of the continents occurred." Finally, in the seventh epoch, man appeared, whose history would extend and complete the history of the earth.

This account was marked off into stages that defined the "epochs." These could be reconstructed by decoding the "archives of nature" and noting and interpreting the "facts" and "monuments" that survived like so many "milestones on the eternal road of time."[39] For the earth, time exists on a different scale from that of human history. Buffon the naturalist was a creditable historian, however, in the way he collected and interpreted the documents from nature's past, the "archives" that were the layers of the earth, the "medals of the past" that were fossil remains, and by reinterpreting religious "traditions." To give meaning to these random signs, he composed a linear account that unfolded from chaotic origin to triumphal achievement.

Epochs of Nature is the work of a historian gathering traces of the past and weaving them into an account—a narrator retracing the movement of beings through the accidents of history marked by successive appearances of the world's beings, and oriented, along the temporal axis, toward the advent of man. It is also the work of a poet[40] playing with the rhythms and harmonies of language. But inherent in *Epochs of Nature* was a desire to understand natural history and to define its laws within a general, rational, Newtonian framework. Buffon's resolutely "actualist" perspective, the place he gave to experimentation (notably the calculations and experiments on the cooling of metals), his rejection of miraculous explanations, the affirmation of the existence of lost species, and of a dynamism of nature in which the earth's history includes that of animal species, all make Buffon an innovator in many respects.

And yet, in its totalizing ambition, *Epochs of Nature* no longer met the scientific demands of natural history when it was written and published. "This system came too late," wrote Jacques Roger in his preface. In the last decades of the eighteenth century, "scientists were more than ever being asked to give precise explanations to certain precise facts." Instead, Buffon had written an ambitious text designed to integrate at once the empirical givens of fossils, geology, and zoology, and to incorporate in a "system" the entire history of the earth from its formation to the advent of man. As a synthesis of knowledge developed since the late seventeenth century, and that took its place in a series of world systems constructed by Descartes and his successors, *Epochs of Nature* constituted the apogee of a genre. At the same time, it marked the decline of that form of discourse known as "theories of the earth."

When Buffon died in 1788, a certain approach to scientific research, a certain style of discourse died with him, as did a certain kind of institutional power. "Whether as intendant of the Jardin du Roi, or lord of Montbard, Buffon only lived in places where he ruled," wrote Roger.[41] Buffon had ruled as a despot in the heart of scientific Europe for more than half a century. The revolutionaries of 1793 clearly recognized this when they abolished the Jardin du Roi among other symbols of the ancien régime.[42] Buffon embodied both the hope of a total understanding of nature and an autocratic system of organizing and producing scientific knowledge. An epoch of science disappeared with him.

A few years later Cuvier would shine new light on *Ohio incognitum*, identifying it as an extinct genus of the Proboscidea family, a kind of cousin to the elephant. The title of a paper that he published in 1806 is "On the great mastodon, an animal very close to the elephant but whose molars are studded with large bumps, whose bones are found in various places on both continents, and especially near the banks of the Ohio in North America, incorrectly called *Mammoth* by the English and by the inhabitants of the United States." In it, Cuvier gave a detailed description of the animal, concluding "that the *great mastodon*, or *Ohio animal*, was very similar to the elephant in tusks and overall bone structure, except for its molars; it very probably had a trunk; it was no taller than an elephant, but somewhat longer, with slightly heavier members and a slimmer stomach."[43]

In writing his study, Cuvier benefited from the debates and observations that had been accumulating for the last half century on both sides of the Atlantic. Among others, he drew on the work of the German naturalist Johann Friedrich Blumenbach. In 1799 Blumenbach made a list of "unknown animals" whose petrified bones had been found buried in the earth. It included "the colossal monster of an earlier world, the Ohio mammoth (*Mammut ohioticum*), whose bones had been dug up in quantity near the Ohio River in

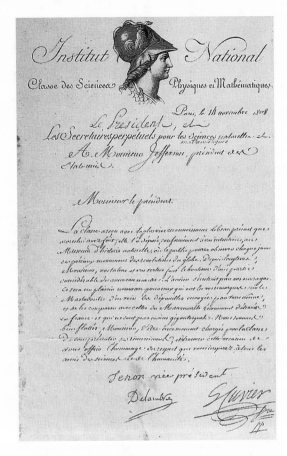

Cuvier's 1806 letter thanking Jefferson for donating an "American mastodon" bone to the Paris Muséum. (American Philosophical Society Library photo, Philadelphia.)

America, and which is notable . . . by the unusual shape of its enormous molar teeth."[44] Blumenbach had read Buffon, Daubenton, and the other naturalists who had studied the remains of the famous Ohio animal and its knobby teeth. He became the first person to give scientific names (in accordance with the rules of Linnean nomenclature) to the Russian mammoth (*Elephas primigenius*) and the North American mammoth (*Mammut ohioticum*). These were paradoxical designations, since *Mammut* refers not to the Russian mammoth, but to the American animal. When Cuvier studied this very same animal in 1806, he noted the breastlike crests on its molars and gave it the name of "mastodon" (from the Greek *mastos* meaning "breast" and *odous*, "tooth"). But Blumenbach's scientific names, which had precedence, have been maintained. This is why the Siberian mammoth has the scientific name *Elephas primigenius*, while the mastodon is still known today in scientific nomenclature as *Mammut*. To complicate things further, it was soon learned that mammoths and mastodons had coexisted during the Quaternary in America and Europe, and that their remains were sometimes found mixed together.

The identification of the "unknown Ohio animal" was made in France, but one can say that the affair of the American "mammoth"—or rather, mastodon—was central to the history of the American nation. "The beginnings of vertebrate paleontology in North America"[45] can only be understood against the backdrop of the conquest of a territory and the birth of a nation by way of its political and cultural institutions.

Nowhere else has the search for and interpretation of paleontological remains been so closely linked to the search for a national identity. It was crucial for Americans to set themselves apart from the colonial power of the countries beyond the Atlantic and also from the Eurocentric portrayals of an inferior, weak, degenerate "savage." Nowhere else was the exploration of the soil's riches and curiosities better linked to the desire for territorial *rootedness* in the face of the twin demands of European colonists and native peoples. Against the prejudices of French and English scientists, American naturalists struggled to uncover the truth about the vestiges of the animals that once occupied their soil. Contrary to Indian "myths," scientific "truths" found their function and meaning against the backdrop of conquest of territory and the creation of the American nation's institutions.

After first winning political independence, Americans gradually acquired their institutional and cultural independence from England and France—though these remained reference points for culture and knowledge. The affair of the American mammoth bolstered the new nation's interest in its own natural and cultural riches, and its capacity to study them within its own institutions. America's first natural history collections and its first museums date from the last decades of the eighteenth century.[46] The first museum of natural history was created in Baltimore in 1808,[47] following the excavation of the nearly complete skeleton of an "American mastodon" by Charles Willson Peale.

The oldest American scientific society, the American Philosophical Society, played a vital role as a locus for collection, research, assembly, and publication of ethnographic and paleontological materials. It also was a sociable gathering place for the greatest scientific minds of the time and helped launch America's first great scientific fossil-hunting expeditions. As the eighteenth century became the nineteenth, true specialists in this domain of science followed the pioneers and amateurs who had done the field research until then. For many decades American paleontology would be practiced by researchers able to organize digs and display collections at their own expense, or who

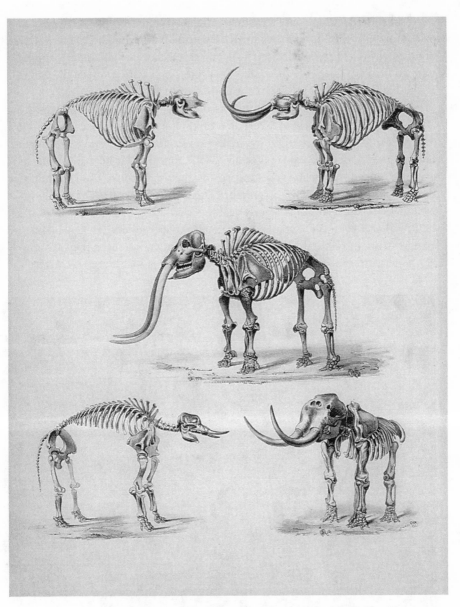

The great diversity of mastodons, illustrated by John Collins Warren, *The Skeleton of the Mastodon Giganteus of North America* (Boston, 1852).

were subsidized by patrons, at a time when France already had a chair of paleontology at the Muséum National d'Histoire Naturelle. But the structure of nineteenth-century French teaching and research was much more strongly state controlled. One cannot speak of a professionalization of the discipline in the United States before the twentieth century.[48]

This eventful episode of the founding of a science at the same time as the "birth of a nation" produced a veritable mythology whose impact is still felt today in American historiography: heroic times, founding legends. One can say that the United States, supposedly "new" and without history, found its true cultural autonomy and roots by proving that its territory's soil contained the remains of a "mammoth" even more gigantic than those of the old continent. The noisy crowds that eagerly visit the Mammoth Site in South Dakota, La Brea Tar Pits in California, or the paleontological collections at the American Museum of Natural History may be only dimly aware that they are paying homage to the remains of the nation's founding ancestors.

Exhumation of a mastodon discovered in 1799 near Newburgh, New York. *Disinterment of the Mastodon* (1806–8) by Charles Willson Peale shows the digging at the site that began in 1801. (Peale Museum photo, Baltimore.)

The Mammoth and the "Revolutions of the Surface of the Globe"

On the 1st of Pluviôse of the year IV in the French revolutionary calendar, the mammoth became the emblematic figure of a science. On that date, January 21, 1796, citizen Georges Cuvier addressed the science section at the Institut de France and read his *Mémoire sur les espèces d'éléphans vivantes et fossiles* (Treatise on the species of elephants, both living and fossil). The speaker boldly declared that the conclusions of this study on elephants would be of the greatest importance for the theory of the earth and could "shed light for us on the obscure and intriguing history of the revolutions of the globe."

The young man who spoke with such confidence and eloquence at the podium had arrived in Paris less than a year earlier, on March 20, 1795. Georges Cuvier was born to a modest family of government clerks in Montbéliard, a small Protestant, French-speaking principality that was part of the duchy of Württemberg. He studied at the Ecole Française at Montbéliard and then at the Karlschule in Stuttgart, where he encountered the works of the great German naturalists of the time, particularly those oriented toward comparative anatomy, geology, and the study of fossils. Cuvier then became a tutor in a French Lutheran family in Caen, the d'Hericy. In 1792, as the Revolution was taking hold, Cuvier was in Normandy engaged in political and scientific activity and studying botany and zoology in the field. There he was noticed by a member of the Académie Française, abbé Henri-Alexandre Tessier, who recommended him to his friends in Paris: Jussieu, Parmentier, and Geoffroy Saint-Hilaire, who was then the director of the Jardin des Plantes.

In the years following the French Revolution, scientific institutions were undergoing major changes. The Institut de France was founded in 1795 by the Convention to take the place of the old academies of the ancien régime. The Muséum National d'Histoire Naturelle, which replaced the former

Jardin du Roi, had come into existence only in 1793. These fledgling institutions were looking for new talent, and in 1795, when he was twenty-six, Cuvier was elected a member of the Académie des Sciences; he would later become its permanent secretary.

In that same year, Cuvier was named an adjunct professor to the chair of animal anatomy at the Ecole du Muséum, whose holder was the zoologist Antoine Mertrud. Cuvier impressed such eminent naturalists as Daubenton and Bernard Lacépède with his culture, intelligence, and talent as an anatomist. He brought the German naturalists' ideas and methods to the practice of comparative vertebrate anatomy. The richness of the zoological collections of the former Jardin du Roi, which had been greatly increased thanks to Buffon's activity and prestige, played an important role in the formation of a new scientific approach and a new discipline.

Cuvier's 1796 treatise[1] was one of the first scientific texts he published; it is also the cornerstone of scientific paleontology. Its purpose was to establish "in an irrefutable way" that at least two species of elephants exist, "which differ by climate, behavior and shape." Naturalists of western Europe had recently learned more about the anatomy of elephants. A few years before, during the winter of 1774, the Dutchman Pieter Camper had dissected the carcass of a young Asian elephant and published the results, with very detailed illustrations.[2] Other dissections of Indian elephants had been carried out in the last decades of the eighteenth century by Perrault, Blumenbach, and Mertrud. And several elephant skeletons were to be found in the collections of the Muséum National d'Histoire Naturelle of Paris, which had recently been enriched by the acquisition of the zoological collections of The Hague's Stadholder following the victory of the young French republic's armies in Holland. In his comparative studies of elephants and the Russian mammoth, Cuvier also was heir to all of the debates of the second half of the eighteenth century in Russia and in Germany, the echoes of which had reached England and France.

Cuvier chose to focus on the "species of elephants, both living and fossil" by first demonstrating that the elephants known today—the Asian and the African elephant—did not belong to the same species, as was then believed. To establish this difference, Cuvier carefully catalogued their distinguishing traits. Asian elephants live in dry, high areas and can be domesticated. But Cape elephants, which live in the humid African plains, are considerably larger and stronger and have never been domesticated. One could also see differences in the anatomy of the skull, the morphology of the teeth, and the shape and length of the tusks. In Asian elephants the forehead is concave and the molars formed by straight, parallel enamel plates. In African elephants

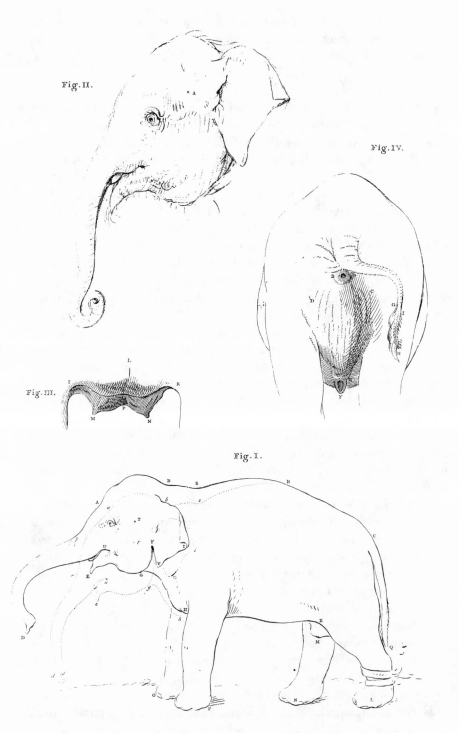

Dissection of an Indian elephant by Pieter Camper during the winter of 1774. *Déscription anatomique d'un éléphant mâle* (Paris, 1802).

the head is shorter, the forehead convex, and the sheets of molar enamel are lozenge shaped. Cuvier argued that these significant differences could not be explained by the influence of climate alone. To claim—as Buffon would have—that these were the effects of "successive degenerations" caused by circumstances would be "to reduce the whole of natural history to nothing, for its object would consist only in variable forms and fleeting types."[3] For Cuvier, the animal world is a fixed one—which is why one can describe it scientifically, determine its laws, and establish its rigorous classifications. One must therefore conclude that there exists today not one, but two distinct elephant species.

Cuvier then came to the main point of his demonstration. In addition to the two living species, a third species exists of which "the originals" are no longer known to us. The second article of this treatise is called "Des espèces d'éléphans perdues" (On lost elephant species). Here Cuvier took up the famous question of the Russian mammoth, which had caused so much talk in Europe since the end of the seventeenth century. "Everyone knows," he wrote, "that a great number of remarkably large bones are found in Siberia, quite close to the surface, and still little altered." People in those countries said that they belong to animals living undergrounds, like moles, and collected and used their "horns," but they were none other than "tusks like those of Elephants, and composed of the same ivory that is used in artwork. . . . More reasonable travelers such as Gmelin and Messerschmidt considered these bones as having come from Elephants."[4] But on close examination, were they indeed those of elephants, like the ones living today? To find out, Cuvier meticulously compared the anatomy of the mammoth's skeleton, shape and length of the tusks, skull shape, and dental anatomy with those of living elephants. A number of differences emerge. The mammoths' molars have "narrower ribbons," which are thinner and more closely packed than those of Indian elephants, and "the lower jaw [is] more obtuse." As to the skull, Cuvier was familiar only with Messerschmidt's drawing, which Breynius published in 1741.[5] But he had noticed in the drawing that the sockets of the tusks were "proportionally twice as long, in relation to the dimensions of the head, as in Indian elephants." And this fact, he wrote, was the source of his intuition. From this anatomical comparison, one must conclude that the Siberian mammoths differed "as to species" from living elephants, although they sufficiently resembled them to be considered members of the same genus. They were *fossil* elephant species, "lost species."

Cuvier's brilliant demonstration was the birth of a new scientific discipline. The naturalist was revealing a new truth, founded on "facts of observation," not myths or legends. The anatomical comparison made evident the existence of several distinct species of elephants, some of which had disappeared from

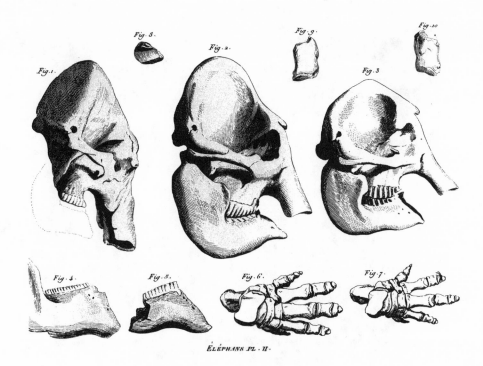

Cuvier's anatomical comparison of the bones of the skull and extremities of the "Russian mammoth" (*left*), the Indian elephant (*center*), and the African elephant (*right*). The plate established the existence of three distinct species of elephants: two living, one extinct. *Recherches sur les ossements fossiles de quadrupèdes,* vol. 2 (1812). Cuvier's representation of the mammoth's skull (*left*) was borrowed from Messerschmidt's drawing (see page 74).

the surface of the globe. And that was Cuvier's essential thesis, the foundation of his work as a paleontologist.

True, the idea of "lost species" was not a new one at the end of the eighteenth century. Buffon accepted it, and Cuvier had no doubt encountered it at the Stuttgart Karlschule in the writings of German naturalists. In the sixth edition of his *Manual of Natural History* (1799),[6] Johann Friedrich Blumenbach suggested that "petrifications" and "fossils" should be classified in terms of the degree of their *resemblance* to animals living today: "It is still more instructive and important to Geogeny [the formation of the earth], to consider Petrifications in a double point of view; viz., first as regards the beds in which they are now found; and secondly, as regards their identity, or mere similarity, or total difference from the organized beings of the existing

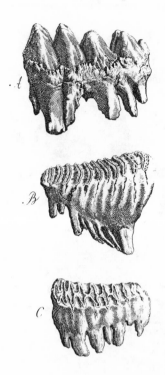

Blumenbach's comparison of molars from animals that were known (C: the African elephant), completely unknown (A: "the animal of the Ohio"), and "doubtful" (B: the mammoth). *A Manual on the Elements of Natural History* (6th ed., 1799).

Creation."[7] Comparing petrifications with "organized beings of the existing Creation" allows one to distinguish, among fossils, petrifications which can be positively determined (*Petrificata superstitorum*), doubtful petrifications, that is, merely resembling creatures at present existing (*Petrificata dubiorum*), and petrifications of perfectly unknown creatures (*Petrificata incognitorum*).

For "doubtful" petrifications, there were some "differing from [others] sometimes by various slight but uniform deviations in the form of particular parts." In this category, Blumenbach in particular included the remains "of the Mammoth of the Old World, a Species of Elephant (*Elephas primigenius*), the suppositious Giant's bones of our ancestors."[8] Those bones were found in Siberia, but also in great quantity in Germany; Blumenbach cited the work of the zoologist Carl Heinrich Merck and the famous elephant skeleton dug up near Burgtonna in 1695.[9]

As for fossil animals that are completely unknown in nature today, these were "petrifications of perfectly unknown creatures of the primitive world, i.e., those not even resembling, much less identical with, any beings at present known."[10] This text was published in 1799, but Blumenbach had entertained the idea of fossil species unknown in nature today, of worlds destroyed by catastrophes, and of stratigraphic successions of "petrified" beings in the layers of the earth as early as 1779. The combination of these ideas was very close to the theories that Cuvier would embrace.

In his 1796 treatise on elephants, Cuvier replaced Blumenbach's doubts with real certainties. The remains of fossil elephants (to which Blumenbach had first given a Latin name in accordance with Linnean nomenclature, *Elephas primigenius*) were not those of animals whose analogy with living elephants is "doubtful," but indeed the vestiges of beings that had completely vanished from the surface of the earth. The very enormity of some of these extinct animals' remains added to the proof of their disappearance: "How then can it be believed that the huge mastodons and gigantic megatheriums, whose bones have been found underground in the two Americas, still live on that continent?" exclaimed Cuvier.

> How can they have escaped the nomadic peoples that ceaselessly move about the continent in all directions, and who themselves recognize that they [the animals] no longer exist? For they have devised a fable about their destruction, saying that they were killed by the Great Spirit in order to prevent the annihilations of the human race. But one can see that this fable was occasioned by the discovery of the bones, like that of the inhabitants of Siberia, who claim that the mammoth lives underground like a mole, and like all those of the ancients, who identified giants' tombs wherever they found elephant bones.[11]

Cuvier's argument was a substantial one, though it probably would not have carried much weight a century or two earlier, when nature was still seen as a Pandora's box, a dark and unknown reservoir of new living shapes, whose marvels Europeans were discovering during distant journeys. In those days one could believe that far-off lands were inhabited by giants, Cyclops, one-legged men, dragons, or salamanders. But by the very end of the Enlightenment, when Cuvier was writing, the earth had been crisscrossed in all directions, and a systematic classifying approach to beings in nature, plants, animals, and humankind had replaced the credulous and sometimes embellished travelers' tales of centuries past.

Not only had living species disappeared, Cuvier argued, but of necessity, all fossil species were *lost* species. Cuvier called fossils "beings from a world before ours . . . beings destroyed by some revolution of the globe."[12] By then, the word "fossil" had finally lost its vague and general meaning of "object buried in the earth" that it had had since the sixteenth century. And the discipline that Blumenbach still called "oryctology"—the science of beings buried in the earth—in 1822 took the name of paleontology[13]—the science of ancient beings that will never again inhabit our world, except through the pen and the pencil of that demiurge, the paleontologist. And this demonstration would find a choice and spectacular subject in the study of elephants.

The fossil species was certainly distinct from living ones. And that was shown even more clearly in 1799 when the botanist Mikhail Ivanovich Adams uncovered and attempted to excavate a mammoth carcass preserved in the frozen earth in Tungusy at the mouth of the Lena River. This revealed the characteristics of the animal that distinguished it from living elephants, but also yielded the key to these elephants' survival in the far north: Adams's mammoth was covered with a woolly coat and was perfectly adapted to the climate that reigns in those northern regions. The discovery of the Lena mammoth turned the elements of the problem upside down while making them much simpler. "One specimen, which Mr. Adams recently extracted from the ice on the Siberian shore, seems to have been covered by two kinds of thick hair, making it possible that this species had lived in cold climates," wrote Cuvier.[14] It was no longer necessary to explain the presence of tropical species in countries that are cold today by hypothesizing a gradual cooling of the earth or a change in the angle of its axis.[15]

But if the mammoths' survival in Siberia had become easily explainable, their disappearance remained to be accounted for. Once the main characters of the story had been found and described, one needed a system of explanation within which these fossil beings could fit; those extinct beings—and their extinction itself—must be shaped into a credible narrative. Cuvier proposed to explain the successive extinction of species by the effect of "revolutions" that had affected "the surface of the globe." The history of the earth was marked by cataclysms that have completely annihilated living beings several times, he wrote. "All these facts, consistent among themselves, and not opposed by any report, seem to me to prove the existence of a world previous to ours, destroyed by some sort of catastrophe."[16]

The catastrophist idea was not new when Cuvier was writing, either. Throughout the eighteenth century, a vision of the earth as a "field of ruins" had haunted people's imagination. Blumenbach and the Swiss geologist Jean-André Deluc had explained "lost species" by successive catastrophes, which had regularly occurred in the history of the earth. For them, the study of fossils illustrated the question of geogeny—that is, of the formation of the earth—"and the various more or less widespread catastrophes to which our earth had been successively exposed."[17] Much earlier, diluvialists like Burnet,

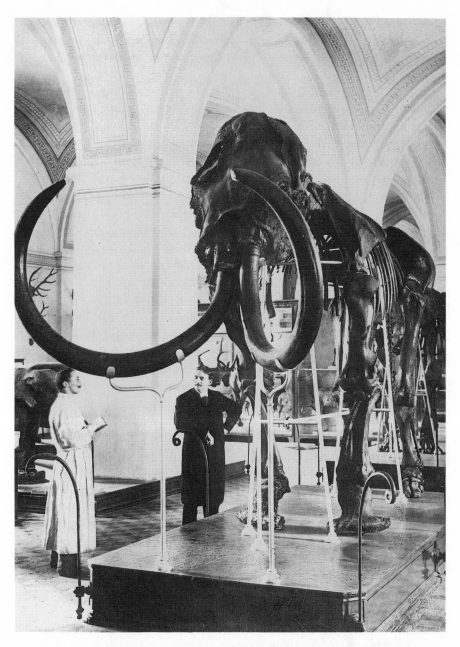

Adams's mammoth in the St. Petersburg Zoological Institute, in a photo taken after 1900. Discovered as "flesh and bones" on the banks of the Lena River in 1799, the specimen could not be excavated until 1804 and the soft tissues were lost. The St. Petersburg Academy of Sciences bought the skeleton (minus the tusks, which had been sold separately) for 8,600 rubles. The mounting of the skeleton was changed several times: the position of the tusks in 1899, and the line of the back around 1950. (St. Petersburg Zoological Institute archive photo.)

Woodward, and Scheuchzer had seen in the jagged and tortured relief of the earth the vestiges of a lost paradise and had tried to mesh their explanations with the Scriptures by introducing sudden and miraculous events modeled on the Flood into the formation of the earth. Others, like John Ray, had envisioned earthquakes, volcanic eruptions, or tidal waves, while maintaining the idea of a short time frame, which fit the biblical chronology. But some materialist thinkers, such as Boulanger or d'Holbach, had also hypothesized great cataclysms. Early in the eighteenth century, this approach contrasted with the notion of a continuous history of the earth determined by gradual causes: a slow retreat of the sea, the gradual formation of layers of the earth or mountains by the accumulation of sediments. That thesis was generally linked with the idea that the history of the globe unfolded over an immense time span. This was the view of amateur naturalist Benoît de Maillet (1656–1738), who estimated the history of the earth at "more than two billion years," based on "the diminishing of the sea" over the ages from the time when it covered the summit of the highest mountains formed within it. These ideas were put forth in his *Telliamed*, a clandestine work printed in 1748 but that had circulated in manuscript form as early as 1720. This was also Buffon's belief at the end of his life, when he thought he had found the key to the history of the earth and its living creatures in the idea that the earth was originally an incandescent globe and was slowly cooling. Lamarck—a disciple of Buffon in many respects and Cuvier's contemporary—had also adopted the thesis of a slow transformation in his *Hydrogéologie* (1802) and in his ideas about transformations within the animal kingdom, which he presented in his *Philosophie zoologique* (1809). All of these "continuist" theses were linked to an "actualist" philosophy, which saw the history of the earth and living beings as a gradual transformation determined by the same causes that are acting in the present.[18]

Cuvier's opinion was completely otherwise, and resolutely catastrophist. In his eyes, the way that the mammoths appeared to have frozen to death and the perfect state of preservation of their remains proved how suddenly these animals had died—and therefore proved the brutality of the causes that led to their extinction. "But whatever this event may have been, it must have been sudden," he wrote.

> The bones and ivory that are so perfectly preserved on the Siberian plains are so only because of the cold that freezes them there, or that in general arrests the action of the elements on them. If this cold had only come slowly and by degrees, these bones—and, with even stronger reason the soft parts in which they are sometimes, even if rarely, still covered—would have

had time to decompose like those found in hot and temperate countries. It would be quite impossible that an entire carcass, such as the one Adams found, could have kept its hairs and skin without decay if it had not immediately been enveloped in the ice that preserved it for us. In this way, all the hypotheses of a gradual cooling of the earth, or of a slow variation in either the inclination or the position of the axis of the globe, collapse by themselves.[19]

The very conditions under which the animal's remains had been preserved pointed to a sudden event that had caused its death. In this way the destiny of the Siberian mammoths became the model for all of those extinct animals (*Megatherium* and American mastodons, the crocodile from Saint Peter's Mount at Maastricht, and the *Paleotherium* found in Montmartre gypsum), the remains of which are found buried in the earth.

For Cuvier, current causes could not explain the sometimes "chaotic," ruinlike aspect of the earth we see today. None of the causes "that are taking place on the surface of the earth today" could have produced "these upheavals, these rendings, these fissures," which could be seen in the layers: rain, freezing and thawing, running water, marine erosion, or volcanism are insufficient to account for this aspect of the globe. The true causes are of such violence and suddenness that they could not be compared to those that we see now acting in nature.

What we find in the earth, said Cuvier, are partial, discontinuous, and scattered remains, sometimes separated by thick layers of sterile sediments. He tried to account for the results of observation in a completely descriptive way. The history he told—that of the sudden disappearance of fauna with the layers that contained them—exactly reflected what he *saw:* a history that was broken, discontinuous, subject to sudden brutal ruptures. It presupposed a *fixist* interpretation of the succession of fossil "worlds." Marine animals, mollusks and vertebrates, then amphibians had been the first to appear, followed by reptiles and mammals, and, finally, humans. This succession, the "thread" of which was "broken" several times, this growing complexity of organisms, took place in a time period that was as short as possible, compressed almost into a succession of instants. The effects of catastrophes, which are both "sudden and massive," did not in any way permit these events to be linked together by the thread of a continuous narrative.

Cuvierian time is the discontinuous time of geological catastrophes; it is also that of the irreversible succession of fauna. Ancient beings succeeded one another and "worlds" replaced each other in a hierarchical gradation. The absence of "intermediary forms" in this discontinuous succession of layers of the earth is one of the major arguments that Cuvier and his disciples used against Lamarck's gradualist transformism. "If species have changed by degrees," he wrote, "one ought to find some traces of these gradual modifications; that one ought to find some intermediate forms between the palaeotherium and present-day species, and that up to now that has not happened at all."[20] This history of the earth marked by cataclysms could be read in the succession of species buried in its layers. The geology of the Paris Basin studied by Cuvier and Alexandre Brongniart[21] became a model for understanding the geological configuration of other regions of the world.

Cuvier claimed to be the founder of a science and is today celebrated as such. The possibility of reconstituting "lost worlds" opened the immense scope of a new field of knowledge. Invoking the model of Newton, who had formulated the law of universal gravity, Cuvier set out to establish the laws for understanding living beings, their structure, and their history.

Cuvier's approach in this regard was close to Auguste Comte's positivism. Science develops relational laws and comes into its own by eliminating imagination, myths, and fables. Against Siberian legends, the myths of the ancients, and American Indian "fables," Cuvier set the cornerstone of a new discipline based on comparative anatomy, which, in his eyes, must replace all the accumulated mythologies and superstitions about fossils. "We are past the time when ignorance could claim that the remains of organized bodies were simply sports of nature created in the bosom of the earth by its creative forces."

The reconstitution of lost species, wrote Cuvier, is made possible by "an almost unknown art . . . a science hardly touched on hitherto, namely that of the laws that govern the coexistence of the forms of the different parts of organisms."[22] Those laws could be deduced from the study and comparative anatomy of living and fossil beings—of which his treatise on elephant species was a brilliant example. The first principle in determining the species to which an animal belongs he called "the correlation of forms in organisms," in conjunction with a functional conception of the organism. "If the intestines of an animal are organized in such a way as to digest only flesh—and fresh flesh—it is also necessary that the jaws be constructed for devouring prey; the claws, for seizing and tearing it; the teeth, for cutting and dividing its flesh."[23] Thanks to this conception of the organism as a functional system, it is possible to "re-create the entire animal" from a fragment of skeleton or a single tooth. "In a word, the form of the tooth entails the form of the condyle; the forms of shoulder blade and the claws, just like the equation of a curve, entail all their properties."[24] By his very vocabulary ("equation," "curve," "entail," "properties"), Cuvier was enunciating his "principle of correlation" as a mathematical theorem. Using the rhetorical device of the synecdoche (in which the part refers to the whole), he presented with great confidence one of the most powerful and enduring images connected with the paleontologists' profession: the spectacular practice of reconstruction, which, starting from a small part, produces a gigantic creature. It is a prodigious alchemy akin to the myth of the phoenix rising from its ashes, or the resurrection of the dead.

More than the *causes* that ruled objects and phenomena, Cuvierian science studied the structural laws that determined the *relations* that linked them together in "closed systems": principles of "correlation" and "subordination of functions," connections between "the history of fossil bones" and "the theory of the earth," relations between animal and plants remains and the "mineral layers which contain them." This twin—and necessarily connected—history of the earth and of living beings, would have an impact on the knowledge of anatomy, "the physical history of the globe," mineralogy, geography, "and even, one may say, the history of mankind."[25]

In that history, quadruped fossils would play a central role. They are better landmarks in time—as stratigraphic indicators—and also in space than the remains of invertebrates or marine animals because they show the limits of submerged continents better than marine animals do. "The nature of the revolutions that have altered the surface of the globe must have had a more thorough effect on terrestrial quadrupeds than on marine animals."[26] Thus the great reptiles of the Secondary (Jurassic) era rose from the earth: gavials, giant tortoises, pterodactyls, iguanodons; the Tertiary pachyderms found in the Montmartre gypsum; paleotheriums, lophiodons, anoplotheriums, anthracotheriums, cheropotamuses; and in the layers of the more recent epoch, the "gigantic pachyderms," rhinoceroses, hippopotamuses, horses, and large ruminants, and such carnivores as the great bear, lion, tiger, and hyena. Among these animals, which he called from the depths of the ages, Cuvier inserted the portrait of

> the elephant called mammouth by the Russians (*Elephas primi-genius* Blumenb.), fifteen to eighteen feet tall, covered with coarse reddish wool, and with long black stiff hairs forming a mane along the back; its huge tusks set in sockets that were longer than those of modern elephants. . . . It has left thousands of its remains from Spain to the shores of Siberia and throughout North America; it was thus widespread on both sides of the ocean, if in fact the ocean then existed where it is today.[27]

The paleontologist, a "new species of antiquarian," had learned "to restore these monuments of past revolutions" and "to ascertain their meaning,"[28] that is, to integrate them into a history. Archaeology, that auxiliary of human history, appeared as the methodological model of this discipline. "If [people] take an interest in following, in the infancy of our [own] species, the almost erased traces of so many extinct nations, they will doubtless find it also in gathering, in the darkness of the earth's infancy, the traces of revolutions previous to the existence of every nation."[29]

Cuvier only sketched the barest outline of the history he was describing, however, and he left many questions unanswered. What were the nature and the rhythm of these catastrophes, these "great and terrible events"? What caused them? Perhaps out of caution, he avoided raising and answering these questions because it would mean directly confronting the relationship between science and religion.

Still, Cuvier suggested that the main causes for the destruction of fauna were certainly due to the movement of water (deluges or flooding). "Living organisms without number have been the victims of these catastrophes. Some were destroyed by deluges, others were left dry when the seabed was suddenly raised."[30] But changes in climate could also have played a role, explaining, for example, the presence of "the bodies of great quadrupeds trapped in the ice" "in the countries of the north." From their appearance, Cuvier deduced that "at the instant these animals died, the country they inhabited became glacial." "This event," which Cuvier called "the last catastrophe," "was sudden, instantaneous, and without any gradation."[31]

Cuvier also remained vague as to the mechanism that determined the appearance of new species. Were *migrations* involved, as he sometimes seemed to suggest? That explanation is difficult to reconcile with the discontinuity of fauna. Or were new creations involved, made possible by a divine intervention? Some of Cuvier's disciples accepted this idea. Alcide d'Orbigny, for example, posited no fewer than twenty-seven "special creations," but Cuvier himself did not write anything on the subject. The idea of a succession during the history of life, of fauna that were stable and perfect in their composition and fixed to the point where only a cataclysm could annihilate them, required some deus ex machina to intervene and to account for their renewal. The paleontologist, however, was less interested in elucidating these processes than in minutely describing the flora and fauna of earlier worlds.[32] Nor did Cuvier give an estimate of the age of the earth or the length of the history of the globe that he was describing. Catastrophes save time, and they represent Cuvier's rejection of the immense period—contrary to biblical chronologies—required by the hypothesis of uniform causes in Lamarck's transformism.

"I know that some naturalists rely a lot on the thousands of centuries that they pile up with a stroke of the pen; but in such matters we can hardly judge what a long time would produce, except by multiplying in thought what a lesser time produces."[33] Did the different parameters that Cuvier brought to his historical construction—fixity of species and his assumption of a basic discontinuity in the history of the earth between present and past causes—translate into an implicit faithfulness to religious dogma? The order of "catastrophic" causes is certainly that of miracle, of which the biblical Flood is a model. When Cuvier referred to certain of these catastrophes, he wrote of "deluges," in the plural. Didn't the medieval church fathers and certain seventeenth- and eighteenth-century natural theologians also speak of deluges, in the plural? Cuvier used the expression "antediluvian" animals, and though he identified the last cataclysm only indirectly with the biblical episode of the Flood—at least in the first version of his text, the 1812 "Discours préliminaire" (Preliminary discourse) that prefaced his *Recherches sur les ossemens fossiles de quadrupèdes* (Researches on fossil quadruped bones)—the revised 1825 edition is much more explicit in this regard.

The original "Discours préliminaire" presented the methodological, philosophical, and narrative framework of Cuvier's anatomical and paleontological studies, which had first been published separately in scientific journals and were collected in four volumes. Rewritten and expanded in 1825, it was widely reprinted as a separate volume, *Discours sur les révolutions de la surface du globe* (Discourses on the revolutions of the surface of the globe). This is the text, much more than the minute studies of comparative anatomy, that was read in the first third of the nineteenth century and that alone is still reprinted today. It is an unusual text in that its readership changes midway. What was originally the preface to a scientific treatise becomes a popular work addressed to the general public. This revised *Discours* includes long sections on ancient traditions and sacred texts alongside geological evidence for the last great catastrophe. There Cuvier was no longer doing anatomy, but comparative mythology. During the hundred-odd pages that he added in the middle of this new version, he examined one after another the most ancient histories not only in the Bible but also Egyptian, Chaldean, and Hindu histories. To date the last cataclysm, Cuvier studied these traditions, and the one he decided was the most reliable of all was—of course—that of the Holy Scripture. This study, which occupies the entire central part of the book, leads to the conclusion that a deluge in fact took place, about five thousand years ago. It is also striking that in this late edition, Cuvier revived the term *"diluvium,"* which until then had been used only in the English version of his treatise[34] and which Reverend William Buckland had used in a framework combining natural theology with Cuverian themes.

By reviving the term *"diluvium"* in the revised edition of the *Discours,*

Cuvier seems to have wanted to maintain a certain ambiguity. The last catastrophe, the one that annihilated the "great pachyderms," the woolly rhinoceroses and mammoths whose remains are found scattered on the surface of the earth, could be identified as the biblical Deluge. "If there is anything that is established in geology," he wrote,

> it is that the surface of our globe has been the victim of a great and sudden revolution, the date of which cannot reach back more than five or six thousand years; that in this revolution the countries in which men and the species of animals now best known previously lived, sank and disappeared; that conversely it laid dry the bed of the previous sea, and made it into the countries that are now inhabited; that since that revolution the small number of individuals spared by it have spread out and reproduced on the land newly laid dry; and that consequently it is only since that time that our societies have resumed a progressive course, that they have formed institutions, erected monuments, collected facts of nature, and combined them into scientific systems.[35]

The "great pachyderms" were destroyed by the last deluge, which marks at least the beginning of human history, if not the rise of humankind. Before that, wrote Cuvier, "everything leads one to believe that the human species did not exist in the country where fossil bones were discovered at the time when the revolutions buried those bones." This is a cautious formulation that left open the possibility—which occurs in the Genesis account (and notably the episode of the Flood and Noah's ark)—that man in those days "might have lived in some limited areas from which they could have repopulated the earth after those terrible events."[36]

Cuvier several times proclaimed the autonomy of science from religion and stressed the fact that speculation on the biblical Deluge distorted the direction of geological research. Yet in the history that he constructed, he in fact agreed with Scripture.[37] On essential points of Christian belief, such as the fixity of species and the intervention of sudden (miraculous) causes, the recent origin of man, the way in which new animal species appear after the revolutions that he described, he maintained a framework compatible with Christian doctrine. From this complicated and ambiguous relation between Cuvier's work and religion, its accounts and dogmas, and the changes of his position in this respect, should we conclude that Cuvier was "a lukewarm hypocrite in religious matters"? Did he align himself with the dogma of the Deluge and the short chronology of the Bible "in order to join the increasing tide of official encouragement of religious orthodoxy in the years after 1802, and hence strengthen his hold on his administrative positions"?[38] Or should one read Cuvier's work both as the first scientific paleontological work but also the last theory of the earth, the ultimate attempt to reconcile scientific knowledge with a literal reading of the biblical narrative? The question remains open, but it would seem that in Cuvier's thinking, biblical history remained as a framework for the history of nature.

Cuvier wanted to create a science of the earth's past and its creatures. But the development of this narrative raised theological questions and religious beliefs, and re-creating a history from the fragmented evidence available to the paleontologist almost inevitably means drawing on fiction and the imagination. Does not every search for an origin encounter dream and myth? Though he embraced the supremacy of positive, rational science, Cuvier was unable to eliminate imagination from his approach. He in turn built a mythical narrative of the origin and history of the earth and living creatures, of the succession of fixed fauna and periodic catastrophes. Did he not himself create more than one account and fable, if not an entire mythology? At a deeper level, doesn't positive faith in science and reason and the ambition of creating an entirely rational science of the history of the living world mean partaking of a myth?

As revised in 1825, the *Discours sur les révolutions de la surface du globe*—by its structure, composition, themes, style, and even its vocabulary—was suited to support a new "mythology" to generate a real fascination for the general public. In the first chapter of his 1831 novel *La Peau de chagrin*, Balzac mixed the fantastical with a sharp and realistic description of Parisian society of the time and paid homage "to the great Cuvier," which is probably that of an entire generation.

> Have you never launched into the immensity of time and space
> as you read the geological writing of Cuvier? Carried by his
> fancy, have you hung as if suspended by a magician's wand over
> the illimitable abyss? . . . Is not Cuvier the great poet of our
> era? Byron has given admirable expression to certain moral con-
> flicts, but our immortal naturalist has reconstructed past worlds
> from a few bleached bones; has rebuilt cities, like Cadmus, with
> monsters' teeth; has animated forests with all the secrets of zool-
> ogy gleaned from a piece of coal; has discovered a giant popula-
> tion from the foot of a mammoth.[39]

The evocation of "revolutions of the surface of the globe" joined a rhetoric of the grandiose and the spectacular suited to create a new mythology of "lost species," cataclysms, and destroyed worlds and gave Cuvier's text the status of a true work of literature. Its chosen object (gigantic quadrupeds, "great pachyderms") and the history it told (the revolutions of the surface of the globe) suggest the grandiose character of the discourse. Cuvier evoked monstrous animals and the suddenness of gigantic cataclysms that annihilated them, which could be read in "the rending, uplifting, and overturning of layers." These ruined landscapes created by "primitive mountains" display "jagged peaks," "fanglike ridges whose crest is torn away," "huge deposits . . . shattered and overturned."[40] Cuvier restored order to the succession of sediments and pebbles that constitute the layers of the earth, but he also read dramas and cosmic tragedies in them. "The masses of debris and rolled stones, which in many areas are found between the solid beds, attest to the force of the movements that these upheavals generated in the body of water."[41] And he concluded, with a tragic lyricism: "Thus life on earth has often been disturbed by terrible events. . . . Living organisms without number have been the victims of these catastrophes. Some were destroyed by deluges, others were left dry when the seabed was suddenly raised. Their races are even finished forever, and all they leave in the world is some debris that is hardly recognizable to the naturalist."[42] "Frightful," "catastrophe," "swallowed or drowned," "suddenly" (see "these great and terrible events," elsewhere) are part of a tragic and deliberately hyperbolic vocabulary designed to impress the reader and to make of the naturalist a new hero and of natural history a new kind of epic. Victor Hugo's *Légende des siècles* was here foreshadowed by the "tragedy of the millennia."

In his account of the origin and history of the world, Cuvier was also building his own myth. He erased his German sources, eliminated his rivals and the naturalists who preceded him in this area, and claimed to have raised the study of fossils to the rank of science by giving it a previously missing rationality. "What I present here comprises only a very small part of the facts that ancient history should embrace. But these facts are important: several of them are decisive, and I hope that the rigorous way in which I have proceeded to identify them will allow them to be considered as points that are definitely fixed," he affirmed.[43] And at the very moment when he declared that he was founding a science and constructing a new kind of history, Cuvier embodied the new figure of a hero of science—the paleontologist, able "to encompass the extent and grandeur of these ancient events" at a glance. Speaking in the first person, he presented himself as the discoverer of unknown worlds attempting to "travel a road on which only a few steps have so far been ventured,"[44] like the demiurge of a new creation who could "burst the limits of time" and "by observations . . . rediscover the history of the world, and the succession of events that preceded the birth of the human species."[45]

The *Discours*, which was first written as the preface of a scientific treatise, as Descartes' *Discours de la méthode*, can be read like a kind of novel whose hero was the paleontologist himself, the re-creator of worlds, capable of re-constructing an animal from one of its claws or a fragment of bone. This was an a posteriori construction where Cuvier proclaimed himself the founder of a science without giving much credit to his masters and predecessors. As he presented his spectacular discoveries, Cuvier built his own image of a found-ing father, an image that has been passed down in the pantheon of the history of science.[46] He invented a new language and adorned his ringing phrases with new dramatic-sounding words, words that were sonorous, poetical, and terrifying (mastodon, *Megatherium*, *Megalonyx*). In France in the first third of the nineteenth century, Cuvier's works and writing created the myth of the paleontologist bringing vanished worlds back to life and raising beings from the past as if at the sound of the resurrection trumpet on Judgment Day.

A psychological explanation (ambition, an individual's imperial will) may not fully account for Cuvier's triumph in this first third of the century. The Cuvierian system was the encounter between an individual and the knowledge and dreams of an epoch. It appeared as a particular construction based on known empirical "facts," but also on collective representations of this turn of the century. During those decades, science succeeded in giving a new, universal dimension to the "universal man" born of the French Revolution: that of time. Man, the crown of Creation, lived in "fertile plains, where the regular course of tranquil waters sustains abundant vegetation."[47] But man's very existence supposes immense and terrifying upheavals that occurred be-fore him. In the "terrible events" of antediluvian epochs, one can almost hear an echo of the most recent but no less terrible revolutionary events that France had recently lived through. The reader is warned at the outset that just as today the world in which we live "is never disturbed unless by the ravages of war or by the oppression of powerful men, . . . nature has had its civil wars and the surface of the globe has been upset by successive revolutions and various catastrophes."[48] In 1825, in the reign of Charles X, recalling the "terrible revolutions" of the past in particularly suggestive terms was probably no accident.

The Cuvierian vision is deeply linked to the "catastrophist" imagination of romanticism, to the post-Revolution "worlds in ruins," and to the post-Empire vanquished titans of history. Curiosity and archaeological research, and the interest taken in vanished worlds, so dominated this beginning of the century, as to spark literary creation. From Jacques Delille's *Trois règnes de la nature* to *Paris avant les hommes* by Félix Boitard,[49] Cuvierian discoveries in France inspired poems and romantic stories set around extinct fauna and generated such striking images as the engravings that illustrate Louis Figuier's book *La Terre avant le Déluge* (The Earth before the Deluge).[50] In the decades around the turn of the century, a focus on ruins and cataclysms flourished, from storms, deluges, and crumbling worlds—from Chateaubriand to Balzac, from the preface of Alfred de Musset's *Confession d'un enfant du siècle* to Jules Michelet's and François Guizot's writings about the Revolution. A distinct taste for macabre ruins and graveyards also appeared in English horror novels. And images of catastrophes and deluges would haunt romantic dreams until late in the century. In England in 1860, Alice in Wonderland was drowning in a sea of her own tears.

Cuvier was indeed the "great poet" of his era. The themes of his work encountered myths that were both traditional (giants, monsters, deluges, catastrophes) and modern (revolutions, the heroic Napoleonic image of the great man, the power of science). It created a dreamlike world full of powerful, mythical images that still survive in popular and scientific representations of paleontology. Separating the imagination from the "pure" domain of science became part of what in 1825 was already called "positivism." That was the image Cuvier wanted to give to his own approach. But today it is possible to understand the relationship of science and myth in another way. The science Cuvier claimed to have founded has in turn created stories and myths, and in some ways is closely connected to representations that are deeply rooted in the imagination and the culture of his time. Cuvier's work can be read at once as scientific and "poetic." It shows the inseparable coexistence of imagination and rationality in the discourse of paleontology at the very moment of its founding. In this way the mammoth, while it was at the origin of a science, was also the hero of a new myth of origin in Cuvier's work.

The Mammoth in Victorian Times

England's soil was full of elephants. Mammoth teeth, tusks, and bones were being discovered all along the west English coast, in Suffolk, in Essex, in the Thames valley, and even under the streets of London. Fossil elephant teeth had been found while digging the great sewer of London "in Charles Street near Waterloo Place, thirty feet deep."[1] "A mammoth's skull containing two tusks of enormous length, as well as grinding teeth, was discovered in 1806, in Kingsland near Hoxton, Middlesex."[2] Teeth, tusks, and bones were also found at Ilford, on the road from London to Harwich, in the sand quarries of Oxford and Abingdon, and in a hundred other locations in Wales, Scotland, and Ireland.

In 1823 Reverend William Buckland, the first lecturer in geology at the University of Oxford, published a work whose Latin title[3]—*Reliquiae Diluvianae; or Observations on the Organic Remains Contained in Caves, Fissures, and Diluvial Gravel, and on Other Geological Phenomena, Attesting to the Action of an Universal Deluge*—uses a vocabulary that refers to the tradition of "natural theology" embraced a century earlier by the English "diluvialists."[4]

Enumerating the fossil remains found in English soil, Buckland stressed the proof that they brought to both Cuvier's theory and the truth of the biblical narrative. He himself found some of those proofs in his digs in the "hyenas cave," or Kirkdale Cave, in Yorkshire, where he found the molars of small elephants mixed with the remains of hyenas, rhinoceroses, hippopotamuses, and large ruminants; at Kent's Hole near Torquay; and again in the famous Goat's Hole Cave near Paviland in Wales, where Buckland discovered the ochre-covered bones of the "antediluvian" "Red Lady of Paviland" associated with the remains of a fossil elephant.

For Reverend Buckland, these fossils buried in caves were proof of the action of the Deluge,[5] and he worked to reconcile Cuvier's thesis with his own

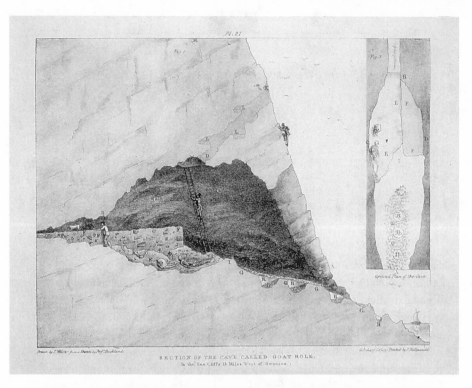

Cross section of the Paviland Cave in Wales. One can make out a fossil elephant skull and a human skeleton (the "Red Lady of Paviland"). William Buckland, *Reliquiae Diluvianae* (1823). (MNHN photo, Bibliothèque centrale, Paris.)

religious position, claiming that the bones of mammoths and other gigantic animals had indeed been buried by a recent "cataclysm," which was none other than Noah's Flood.[6]

Buckland was an admirer of Cuvier's work, which has been called "the scientific bible of catastrophism,"[7] and wanted to carry on his thinking. But where Cuvier had often remained vague as to the concordance between his thesis and religious faith, Buckland's aim was clearly theological. Geology and fossils confirmed the Genesis account. In this way, "descriptive sciences found themselves once again enrolled in the service of natural theology"[8] in England in the first decades of the century.

Those concepts, however, would soon become obsolete. With Cuvier's death in 1832, the Cuvierian representation of the history of the earth and living beings remained triumphant only in appearance. Cuvier's ideas had become dogma at the Académie des Sciences in Paris; Buckland in England and Louis Agassiz in Switzerland, then in the United States until his death

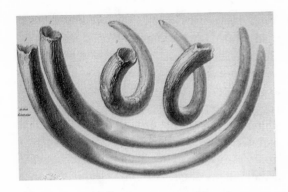

Buckland interpreted the remains of Alaskan fossil elephants as those of animals destroyed by the Deluge. This plate illustrates his appendix to Captain Frederick Beechey's *Narrative of a Voyage to the Pacific and the Beering's Strait* (1831), in which Buckland published the results of his anatomical and geological examination. (Photo E. Rasmussen Library of Fairbanks, Alaska.)

in 1873, defended the fixity of species and of catastrophism; and even after 1860, the myth of Cuvier's infallibility persisted. But his thinking was criticized in many respects. The controversies that had developed during the first decades of the century and from which Cuvier had emerged victorious would be mainly resolved at midcentury by the triumph of his intellectual adversaries. In England Lyell's uniformitarian geology, Geoffroy's conception of a "plan of organization of animals," and the idea of a transformation of living beings set forth by Geoffroy and Lamarck would be clearly developed and considerably reworked by Darwin. The generation of English paleontologists who saw Lyell's *Principles of Geology* published in 1830 would break with Cuvier's catastrophism and Buckland's natural theology. And while most of the discoveries and the battles over the antiquity of man occurred in France, England was the place where the first recognition of fossil man would come in 1859, based on the remains found in the Abbeville "*diluvium*" by Jacques Boucher de Perthes.

It would be up to the generation of naturalists between 1830 and 1860 to try to resolve the contradictions between fixism and evolution, discontinuity and plan of creation, catastrophes and uniformity of geological and biological change. At stake in these scientific positions would be both the limits of scientific knowledge and the ideological and sociopolitical mutations of society.

From the beginning of the century, the mammoth was involved in a history that was both that of the earth and of the living world. In 1830 the problem was not only to prove that living organisms had a history. It also was to understand the logic and nature of their evolution, and to know and to account for the fossil world in all its diversity. Between 1830 and 1860, the single "fossil elephant" species that Cuvier had defined (the mammoth, *Elephas primigenius*) literally exploded into a multitude of new species. The growing number of paleontological discoveries and the spread of new approaches and new modes of classification made it necessary to rethink the principles of replacement of fossil fauna that Cuvier had proposed. Certain forms appeared to be "intermediate," and this raised the burning question of the evolution of species. Was simple succession involved or transformation? Continuity or discontinuity? What was the nature of the processes at work in this history? Were its causes physical or metaphysical? The multiplicity of questions and ideas that characterized these decades would culminate with the publication of Darwin's *On the Origin of Species* in 1859.

New ways of thinking were also the product of the spread of empirical discoveries. This period in England coincided with the apogee of the industrial era, an age of coal and mining, and the triumph of English colonialism. By making the extensive collecting of fossils possible, these two factors played an indirect but essential role in the rise of paleontology.[9] By the middle of the century, the discipline was a shining beacon, attracting young and brilliant minds.

From 1840 on, Richard Owen was a major scientific authority in England. He was a member and later professor of physiology at the Royal College of Surgeons, a lecturer at the School of Mines, and the superintendent of the Bloomsbury natural history collections. Calling himself Cuvier's heir, Owen went so far as to proclaim himself the "Newton" of paleontology who had been announced by the *Discours sur les révolutions de la surface du globe*. In 1858 Owen was elected president of the British Association of Leeds, and in 1863 he founded the Museum of Natural History of London. During his entire life he pursued an enormous project of describing and studying fossils, "which can only be compared," wrote Thomas H. Huxley, "to Cuvier's *Ossements fossiles*."[10] His study of the fossil vertebrates of England, published in 1846 in four hefty volumes, was completed by the publication of three volumes of comparative anatomy, *On the Anatomy of Vertebrates*,[11] in 1866–68 and by several theoretical works of "Philosophical Anatomy."

Like Cuvier, Owen was a man of the laboratory, who studied fossil specimens preserved in museums or private collections. Also like Cuvier, he was a specialist in comparative anatomy. But Owen was a disciple of Lyell and rejected catastrophes in favor of a uniformitarian vision of geological change. In his "Philosophical Anatomy,"[12] Owen tried to reconcile the opposing theses of the "two great lights of the French school," Georges Cuvier and Etienne Geoffroy Saint-Hilaire. The task was to reconcile the Cuvierian conception of the functional unity of the organism with Geoffroy's notion of a transformation of species and a unique morphological plan of organization for the entire animal kingdom.

Owen believed in the transformation of species. He had read Benoît de Maillet and Lamarck, in whom he showed a certain interest. He was familiar with Darwin's work, though he rejected his ideas on evolutionary processes. In his own view of the "derivation" of species, Owen drew neither on Lamarck's habit and will nor on Darwin's mechanism of natural selection. Instead, he believed in an "innate tendency" of beings to metamorphose during the course of the history of species, though he was unable to define the causes of this "derivation."

> To what natural laws or secondary causes the orderly succession and progression of such organic phenomena may have been committed we as yet are ignorant. . . . But [Nature] has advanced with slow and stately steps, guided by the archetypal light, amidst the wreck of worlds, from the first embodiment of the Vertebrate idea, under its old Ichthyic vestment, until it became arrayed in the glorious garb of the Human form.[13]

Owen believed that fauna appeared successively through the action of the continuous creative power of nature. This led Owen to recognize "the continuous operation of natural law, or secondary cause; and that, not only successively, but progressively."[14] He wrote: "The creation of each animal class, Reptiles, Birds, Quadrupeds, has been successive and continuous since the most ancient times."[15] This progressive creation of species that were always well adapted to changing environments could only be understood through the intervention of an "intelligence and foresight" of a divine power. This approach to nature was both aesthetic and theological, blending predestination and transcendental morphology, the representation of a "harmony of nature," continuity, and fixity of species. "Derivation sees . . . a manifestation of creative power in the variety and beauty of the results."[16]

In an address to the British Association for the Advancement of Science in 1846, Owen defined the notion of "archetype," a type that would account for the unity and the diversity of beings. He drew explicitly on Plato: "What animal served as a model to He who created this great animal, the World? . . . The animal that served as an Archetype is that which contains in itself all the other animals as its constituent parts."[17] A "plan," a "model," a unique "idea" dominates the anatomical organization of all vertebrates. This ideal exemplar, which is at the base of the organization of the animals, can only be reconstituted through homologies that can be recognized by their organization, such as "the numerous and beautiful evidences of unity of plan which the structures of the locomotive members have disclosed."[18]

One example that Owen particularly stressed was the lamellar tooth of the mammoth and the elephant, which illustrates both the possibility of a very extensive modification of a structure that is homologous among all vertebrates (the tooth) and the perfect adaptation of an organ to its function.[19] He also stressed the particular way in which molars are formed and replaced horizontally in the mammoth and in living elephants.[20]

In his 1846 *History of British Fossil Mammals and Birds*, Owen turned to the study of fossil elephants, as Cuvier and Buckland had. In the Proboscidea family, Cuvier had distinguished two fossil genera, according to the anatomy of their teeth. Among mastodons, the bumpy crown consisted of tubercules arranged in transverse crests separated by "valleys." The genus *Elephas*, represented by the mammoth and the living elephants, was defined by its particularly complex dental anatomy—a tooth consisting of many thin transverse enamel plates with cement-filled intervals. According to Cuvier, the first genus (*Mastodon*) consisted of "three or four species,"[21] whereas the genus *Elephas* consisted of a single fossil species, the mammoth (*Elephas primigenius*). But when Cuvier defined this unique species, Owen noted, he did not have direct access to the fossils being studied. "The skull of the Siberian mammoth was known to Cuvier only through the drawing of five specimens, none of which was accurate." Nonetheless, through the last edition of *Ossements fossiles* published during his lifetime, in 1825, Cuvier's determination of a single species of fossil elephants was accepted as gospel. Wrote Owen: "Despite the fact that the teeth of fossil elephants from southern Europe generally display longer and fewer plates and thicker enamel than the typical mammoth forms found in Siberia, Cuvier attached little importance to these differences, to the extent that the teeth were similar in other ways; and he classified the whole in a single group, *Elephas primigenius*."[22]

But others thought they recognized new species among those supposed mammoth remains. In 1808 the Italian amateur paleontologist Filippo Nesti identified a new species in fossil remains found in the Arno valley in Tuscany and baptized it *Elephas meridionalis*.[23] In 1821 August Goldfuss described another fossil species, *Elephas priscus*, as the ancestor of African elephants. But Cuvier refuted these different descriptions, which were based on shaky evidence and unconvincing fossils. He showed that the jaw Nesti described was in fact that of a mastodon, and that Goldfuss's fossil was a fake. In the last edition of *Ossements fossiles*, Cuvier maintained his position that there was but one fossil elephant species.

Richard Owen in turn examined and discussed the remains of English fossil elephants at length and attempted to conclude, as Cuvier had, with the specific unity of European forms. According to Owen, all the remains of fossil elephants found in England must be classified in a single species, *Elephas primigenius*. Yet Owen admitted the existence of varieties, distinguished by significant differences in the dental structure. The molars of some varieties displayed thinner or thicker ribbons of enamel, and they did not all have the same number of enamel striations. To explain these differences, he suggested the possibility of a gradation between these different varieties: "Some paleontologists have viewed these differences as indications of distinct species of *Elephas*. But the vast number of grinders of the Mammoth from British strata which have been in my hands in the course of the last three years, have presented so many intermediate gradations, in the number of plates . . . that I have not been able to draw a well-defined line between the thin-plated and the thick-plated varieties of molar teeth."[24]

Owen therefore refused to see several species where Cuvier had mentioned only one and preferred to conclude that these variations were due to age or to "the latitude of variety to which the highly complex molars of the *Elephas primigenius* were subject." To his mind, this diversity affected only the teeth, and not the skeleton as a whole. So he concluded, as Cuvier had, that there was but one species of fossil elephant.

> And this conclusion harmonizes with the laws of the geographical distribution of existing species of Elephant today. Throughout the whole continent of Africa but one species of Elephant has been recognized. A second species of Elephant is spread over the South of Asia and in some of the adjacent islands. . . . If, on the other hand, the observed varieties in the dentition of the Mammoth are to be interpreted, as Parkinson, Nesti, Croizet, V. Meyer, and others have done, we must be prepared to admit not merely three, but six or more distinct species of gigantic Mammoths to have roamed the primeval swamps and forests of England.[25]

For Owen, even more than for Cuvier, accepting the idea of a unique species in the fossil genus *Elephas* stemmed less from ignorance than from a philosophical a priori, the desire to believe in a "harmony of nature" in the distribution of beings. In the case of elephants, this meant a single species corresponding to a given geographical region. But the question of variability had been raised, and Owen himself spoke of "transitional forms" in describing the remains of certain English fossil elephants. Moreover, he had raised the possibility, which he thought improbable, of enumerating as many as six fossil elephant species! This idea—a nightmare for a Cuvierian paleontologist—

would be given shape in the work of another English naturalist, Hugh Falconer.

The year 1830, which was so rich in paleontological debates, was also the one in which twenty-two-year-old Hugh Falconer boarded ship for India, to assume his post as surgeon of the East India Company's outpost in Bengal. Probably the most brilliant representative of English colonial paleontology, Falconer embodied a new kind of researcher: the field scientist. He specialized in the study of the Proboscidea (literally, "animals with trunks") family. During his many digs in the famous Indian Siwalik deposits, he studied and named several new species of fossil elephants and mastodons.[26]

Unlike Owen, Falconer was neither a thinker nor a theoretician. In fact, he seemed a somewhat marginal figure in relation to the scientific establishment of his day. Falconer spent most of this career as a doctor at the military base at Meerut, in India's northwest provinces. In 1832 he also was named director of the botanical gardens at Suharunpoor and had a double career as a doctor and a naturalist for twenty-six years. He was a botanist (he is responsible for the introduction of Chinese tea in India), a paleontologist, and a geologist. He discovered one of the world's richest deposits in Tertiary mammals in the Siwalik Hills at the foot of the Himalayas. He determined in 1831 that the Siwaliks were Tertiary formations and concluded that "the remains of Mastodons and other large extinct mammalia would be found either in the gravels or in other deposits occupying the same position in some part of the range," a hypothesis confirmed by Indian legends of the region, which spoke of the existence "of the bones of giants found in the hills."[27] And it was precisely as the "remains of giants destroyed by the fierce Rasuchundra" that a fossil elephant tooth and tusk were brought to him on December 1, 1834. It would be the prelude to a fantastic collection of extinct vertebrates.

The subtropical mammalian fossil fauna of the Siwaliks was exuberantly rich and luxuriant; he wrote:

> What a glorious privilege it would be, could we live back—
> if only for an instant—into those ancient times when the ex-
> tinct animals peopled the earth! To see them all congregated
> together in one grand natural menagerie—these Mastodons
> and Elephants, so numerous in species, toiling their ponder-
> ous forms trumpeting their march in countless herds through
> the swamps and reedy forests! . . .
>
> Assuredly it would be a heart-stirring sight to behold! But al-
> though we may not actually enjoy the effect of the living pag-
> eant, a still higher order of privilege is vouchsafed to us. We

have only to light the torch of philosophy, to seize the clue of induction and, like the prophet Ezekiel in the vision, to proceed into the valley of death, where the graves open before us and render forth their contents; the dry and fragmented bones run together, each bone to his bone; the sinews are laid over, the flesh is brought on, the skin covers all, and the past existence—*to mind's eye*—starts again into being. . . .[28]

Indeed, the ancient fauna of the Siwaliks was extraordinarily rich. On Falconer's very first day there, he gathered "three hundred specimens of fossil animals" in six hours.[29] On his first return trip to London in 1842, he brought from Bengal forty-eight trunks filled with five tons of fossils, which would provide the material for many publications. Falconer's *Fauna antiqua Sivalensis*, published between 1843 and 1847, would remain unfinished. After a second stay in India from 1852 until his death in 1865, several important monographs on Tertiary mammals were proof of his intense scientific activity.

Falconer did not believe in Buckland's natural theology or Cuvierian catastrophism. He admired Cuvier's work and sometimes seemed to embrace Owen's, Blainville's or Lamarck's ideas, but was also critical of them. A few letters published by Charles Murchison, the editor of his collected works, mention theology, but only to clearly separate the world of reason from that of faith.[30] In Falconer's work there is no grand philosophical synthesis, no vast perspectives on the entire living world, but instead a detailed effort to scrupulously follow the road of facts and the logic of forms. Falconer's work is above all one of description and classification—though these are not neutral, purely mechanical operations. Before they become a reflection of reality, they are driven by the will to explain. And they can bring about a new vision through the image of reality that they deliver.

In the middle of the nineteenth century, the classification of "knowledge of the true order of nature" remained an end in itself,[31] but it was also essential to geology. "It is of the highest importance to Geology," Falconer wrote, "that each mammal found in the fossil state should be defined as regards, 1stly, its specific distinctness, and 2ndly, its range of existence geographically and in time, with as much exactitude as the available materials and the state of our knowledge at the time will admit."[32]

A few general notes in the preface to his *Fauna antiqua Sivalensis* give us clues to Falconer's thinking. Along with the extraordinary diversity of fossil forms in the Siwaliks, he noticed the very great similarity of that fauna with fauna today. But while one encounters the same families in living genera, they are much more poorly represented. "In the Pachydermata, which in con-

"The elephant triumphant over the turtle, supporting the world and deciphering the mysteries of *Fauna antiqua Sivalensis.*" Pencil drawing by Edward Forbes on one of Hugh Falconer's field notebooks. (Published by Murchison, in Falconer, *Palaeontological Memoirs,* 1868.)

tinental India are now restricted to four genera and five or six species, there were then *twice* the number of genera and about *five times* the number of species. Of the Proboscidean Pachydermata alone, including the elephant and mastodon, there were as many Siwalik species as are now comprised in the whole order of Pachydermata in India."[33]

Animals living today appear almost as a relic, a sample of that once-flourishing fauna, as if the animal population had diminished in number and in variety in the course of its history. Falconer concluded: "The era of the great force and development of the vertebrate animals of India has gone, and . . . what we now see as our contemporaries are, as it were, but a ragged remnant representation of the rich garment of life with which the continent was formerly clothed."

At that date (1853) Falconer seemed to be borrowing Blainville's idea of the impoverishment of the animal world. For Henri Ducrotay de Blainville— a French anatomist and paleontologist and a critic of Cuvierian catastrophism—all species, though they were created at the same time, have disappeared successively. Fossil forms that are discovered gradually fill the gaps in the animal series while proving their very ancient existence. The number of species decreases as the living world becomes impoverished. Fossil remains found in abundance in the layers of the earth—for example, those of mammoths and mastodons—are those of species that have disappeared, and all are doomed to extinction. But Falconer did not believe, as Blainville did, in the simultaneous creation of living beings. His view of the history of the living world reveals continuities, probably linked to climatic adaptation. Did he believe, like Lamarck and the transformists, that the scale of beings revealed a progressive transformation over the course of time? Or, like Owen, that a continual creation worked to modify an archetype during the history of the living world? Falconer remained cautious on this point.

Falconer specialized in the study of Proboscidea, as their remains were found in very great quantity among the fossils of the Siwaliks. Many of these "elephant" remains were known in Europe, and since Cuvier's time, fossil elephants had pride of place in the study of vertebrate paleontology. But around 1850 they were starting to raise new and difficult questions.

"The remains of either of the Proboscidean genera, *Dinotherium*, *Mastodon*, and *Elephas*, abound in all the Tertiary formations of Europe, Asia, and America, from the Miocene up to the post-Pliocene," wrote Falconer in 1857. "They have been the subject of a vast number of observations, while it is hardly possible to conceive anything more unsettled and opposed than generally received opinions respecting the species and their nomenclature in the standard works which are of the greatest authority on the subject."[34] In 1845 the French geologist Etienne d'Archiac had listed the localities in which mammoth remains had been found: from the British Isles through the entire temperate zone of Europe and Asia, from the shores of the Arctic Sea to the cliffs of the Bering Strait, and in all of North America as far as Mexico. If one adds the Indian forms, this would lead one to "[assign] at least half of the habitable globe for the pasture ground of the Mammoth," Falconer wrote.

Moreover, the mammoth had apparently lived during an immense span of time, from the beginnings of the Pliocene to beyond the end of the Pleistocene, since "it is only yesterday, as it were, in relation to the human epoch, that the last remnants were exterminated and frozen up in the perennial ice-cliffs of the Arctic Circle."[35] If one recognizes this great distribution across space and time, the mammoth must therefore have survived enormous geological catastrophes during this vast period. So Cuvier's thesis of a single species of fossil elephants contradicts the Cuvierian system itself.

Falconer considered that the species *Elephas primigenius* had been ascribed an extension in space and time that was far too great, being unmatched in the whole class of mammals, fossil or living. A mistaken conception, if not a dangerous one. "Nothing has contributed more to retard the progress of this section of geology in Britain than the generally accepted belief in the specific unity of the mammoth."[36] The image traditionally associated with the name of the woolly mammoth was that of the frozen flesh and bones of the specimen found at the mouth of the Lena River, which suggested the idea of glacial deposits. Cuvier and, after him, his disciple Alcide d'Orbigny—the founder of stratigraphic paleontology—had shown that each layer is identified by the fauna that characterizes it. In which case the presence of fossil elephants was dangerously affecting the perception of the period under consideration, because of a foreordained conclusion. The supposed very vast distribution of mammoth remains was leading to confusion in the distinctions between epochs.

In his 1847 *Fauna antiqua Sivalensis*, Falconer turned to the description of the new species of fossil elephants that Cuvier and Owen had rejected (*Elephas priscus, Elephas meridionalis, Elephas antiquus*) and showed that they were found in Britain in the Pliocene and Pleistocene periods. Moreover, he picked up the description of an Indian form in which "the characteristics that Cuvier had attributed to each of the genera were combined"[37] and that its discoverer had suggestively named *Mastodon elephantoides*, because its dental characteristics seemed (as its name suggests) to represent a morphological transition between mastodons and elephants. Owen himself had called this form "transitional." In Falconer's eyes this discovery cast doubt on the traditional definition of the genera of Proboscidea. On the one hand, the distinction of *Mastodon* and *Elephas* genera was being questioned, as was the determination of fossil elephant species, on the other. How many were there? What was their true extension in time and space? Falconer laid a new foundation for classification of elephants through anatomical study and came up with a key that would answer both questions at once.

Studying the specimens he had gathered in the Siwaliks and the paleontological collections in museums all over Europe, Falconer spent fifteen years trying to shed light on the classification of fossil Proboscidea. His treatise "On the Species of Mastodon and Elephant Occurring in the Fossil State in Great Britain," which he read before the Geological Society of London in 1857, is an answer to Owen's chapter on the *Elephas primigenius* published in *British Fossil Mammals and Birds* in 1846. Basing his demonstration on dental anatomy, Falconer undertook to show that "several European fossil species, belonging to two distinct subgenera, have been generally confounded under the name of *Elephas primigenius*, and that these species are susceptible of being discriminated not on mere trivial or uncertain, but upon broad and well-founded distinctions."[38]

In Proboscidea, six teeth appear successively on each half jaw: three milk molars and three permanent molars. Falconer noted that in certain teeth of the dental series, known as "intermediate teeth"—the last milk molar and the two first permanent molars—the crown is divided into a determined number of enamel crests. This "crest formula" allows one to distinguish mastodons from elephants, by the fact that among the former the division of the molar in three or four crests is quite "isomeric," that is, it is repeated in three "intermediate" molars—whereas among the latter, the crests and intermediate teeth, instead of being limited to three or four, number six or more and they are not isomeric. This classification principle has the advantage of also allowing the determination of subgenera. Among mastodons this produces the trilophodonts (three crests), the tetralophodonts (four crests), and the pentalophodonts (five crests); among elephants, the stegodonts (bumpy, roof-shaped crests), the loxodonts (lozenge-shaped crests), and true elephants (*Euelephas*) (lamellar crown).[39]

Among the English fossils that Owen had identified as those of mammoths, Falconer was thus able to enumerate, in addition to *Elephas primigenius*, three distinct species—*Elephas priscus, Elephas antiquus,* and *Elephas meridionalis*—belonging to two distinct subgenera: the loxodonts and the true elephants, *Euelephas*, differentiated by the shape, number, and distribution of enamel crests on the dental crown. The distribution of species of fossil Proboscidea that Falconer presented in 1857 gave a new picture of the distribution and the history of the Proboscidea genera and species, in which certain groups are presented as "intermediate." The pentalophodont mastodons mark the transition between the tetralophodont mastodons and the genus *Elephas*. Within the genus *Elephas*, the stegodonts represent an intermediate form between mastodons and elephants. The loxodonts are the transition between stegodonts and true elephants, with this latter group representing "the most aberrant type of the ordinary type of Pachydermata." Thus through the study of dental anatomy, a treelike schema emerges that represents a transformation from the mastodons' bumpy tooth to the "most aberrant type" of the mammoths' lamellar molar, within which "the stegodonts constitute the intermediate group of the Proboscidea from which the other species diverge through their dental characters on the one side into the Mastodons, and on the other into the typical Elephants."[40]

Falconer's paleontological work can be called a simple classification of Proboscidea species and genera, but in breaking the narrow framework of Cuvierian classification, this impressive work actually led to a new vision of the fossil world. In undertaking to systematically account for the diversity and complexity of an important family of mammals, it renewed the way the succession of animal species was represented. What Falconer brought to light in his classification of fossil Proboscidea, through the multiplicity of forms, was a gradation that is not only from the most "primitive" to most "evolved," but that gave the image of a kind of bush. By the continuities it showed in determining the "intermediate forms" and by the diversity of forms it

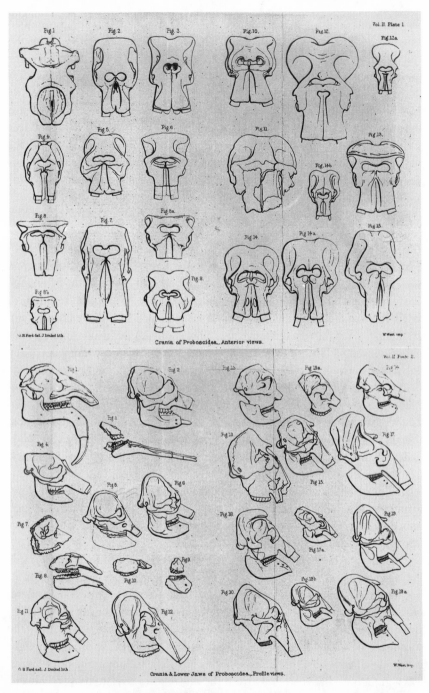

The anatomical diversity of the Proboscidea, according to Falconer.
The skulls representing the different species are side by side, but not
arranged in an evolutionary tree. *Palaeontological Memoirs*, vol. 2
(1868). (Harvard Museum of Comparative Zoology Library photo.)

describes, this classification implicitly told a new history of living beings and raised the question of the modality of their evolution.

Thus by 1857, two years before the publication of Darwin's *Origin of Species*, the mammoth had exploded into a multiplicity of species distinguished by their anatomy and by their range in time and space. There was no longer a single species of fossil elephant—as Cuvier, Buckland, and Owen had claimed—but a respectable number of different species distributed in three subgenera, organized in a continual succession, and qualified as "intermediates" with regard to each other.

Like many other scientists of his time, Falconer often spoke of "intermediate" forms, of transitions. This somewhat ambiguous notion of "intermediates"—because it raises the problem of the continuity of the living world and of the very nature of the history of living beings—would fuel debates and stimulate the thinking of many naturalists in France and England. Cuvier had refused to accept the idea of intermediary species, precisely because he grasped the transformist implications of such a notion. Accepting the existence of intermediary species would mean opening the door to the Lamarckian representation of a continuous filiation of species.

In fact, those who thought about intermediates were not all transformists, and it was possible to accept the notion while remaining within the framework of a living world that was not subject to change—by conceiving the ordering of species as a simple succession, as Blainville did, for example. But even then it was but a short step from the representation of "fixed" intermediate species to that of evolution. All that was required was to graft the dynamic of transformation onto the fixed, treelike representation of a succession of living forms to transform the spatial schema of a branching into a temporal representation of a geneaology.[41]

Living species have evolved over the course of an immense history. This thesis, implicit in the works of a number of paleontologists between 1830 and 1860, was openly propounded in 1859 when Darwin published his great work, *On the Origin of Species by Means of Natural Selection*. According to Darwin, species are transformed and the principle of natural selection explains their birth and their extinction. In the huge quantity and diversity of beings that are born and multiply, the "struggle for survival" leads to the consequence that those creatures best adapted to their environment survive and reproduce.

In the mid-nineteenth century, the question of "intermediate forms" was at the heart of many debates and challenges to evolutionary thinking. If, as Darwin thought, "intermediate gradations" are necessarily absent from nature today, then the fossil world, which represents the remains of extinct animals, should reveal the ancestral forms and allow us to re-create the whole fabric of beings that lived in the past. But these "past" intermediate forms, destroyed in the struggle for life, had not yet been brought to light by paleontology. And that absence constituted a major argument of those—including Agassiz, Pictet, Sedgwick, Barrande, and Forbes—who believed in the fixity of species.

Darwin gave two reasons for that absence. The first is the fragility of fossil vestiges. The remains of living creatures fossilize poorly, and their conservation is rare and accidental. "We ought only to look for a few links, some more closely, some more distantly related to each other,"[42] because the archives of the earth are full of gaps, and these pieces of evidence are extremely rare. This is why fossils occupy a place that is both central and marginal in Darwin's work. The matter of fossils occupies chapters 9 and 10 in a book of fifteen chapters. And Darwin listed the conclusions one can draw from the study of fossils in negative terms, as reflected in the title of chapter 9, "On the Imperfection of the Geological Record."[43] Darwin mentioned the fossil evidence only to deplore "the poverty of our paleontological collections." "Now let us turn to our richest geological museums and what a paltry display we behold!" he exclaimed. "Very many fossil species are known and named from single and often broken specimens, or from a few specimens collected on some one spot." A second reason is inherent in the very process of evolution. We sometimes imagine evolutionary continuity in a simplified way. When we look at two species, we expect to find "forms *directly* intermediate between them." But this is a mistake, and the fossil evidence does not satisfy the expectation. A species derived from an ancestor is necessarily different from, because it is precisely through this modification that it has manifested a better aptitude to survive than its vanished ancestor. "We should be unable to recognise the parent-form of any two or more species, even if we closely compared the structure of the parent with that of its modified descendants, unless at the same time we had a nearly perfect chain of the intermediate links."[44]

In the *Origin of Species*, Darwin thus indulged in a complex and subtle meditation on the question of "intermediate forms." While justifying the difficulty of revealing these "missing links" in 1859, he urged paleontologists to go looking for them. A difficult quest, because in evolution as Darwin conceived of it, both a gradual and progressive transformation were at work, as well as a "divergent tendency" by which the ancestor differs from its descendants.

After 1860 many paleontologists worked to find the fossil traces of these "missing links" not only to study the evolution of given groups (Equidae, Proboscidea), but also to examine the most spectacular phases of evolution:

the move to dry land, the origin of tetrapod vertebrates, of mammals, birds, and man. Extinct species of Tertiary monkeys had already been discovered in India, America, and Europe. In 1863 Richard Owen studied the remains that had been found in the quarries of Solnhofen in 1861 of "that strange bird, the *Archaeopteryx*, with a long, lizard-like tail, bearing a pair of feathers on each joint, and with its wings furnished with two free claws," which proved, in Darwin's eyes, the reptilian origin of birds. And Thomas Henry Huxley, a fervent disciple of Darwin, had studied the *Compsognathus*, which according to him marked the transition between dinosaurs and birds.

Hugh Falconer, however, was not part of the Darwinian young guard. Despite the sympathy he felt for Darwin and his work, he did not adhere to his way of thinking unreservedly. He stated his position vis-à-vis Darwin in 1863 this way: "Having long enjoyed the privilege of intimate intercourse with Charles Darwin, I have been for many years familiar with the gradual development of his ideas on the origin of species, and I have been included by him in the category of those who have vehemently maintained the persistence of specific characters."[45]

Despite his insistence on the existence of "intermediary" groups, Falconer affirmed that they were not enough to prove evolution in the terms that Darwin had defined it, because no "sign of transition" can really be detected in the anatomy of animals. In order for the mechanism of evolution "by means of natural selection" to be proven in the succession of fossil species, "variations" susceptible to the action of natural selection would have to be discernible within the species themselves—and the examination of fossils yields no such thing.

Once again, Falconer based his demonstration on the study of fossil elephants. Four successive species of elephants have existed in Europe, for example. "Do the successive Elephants, occurring in the strata, show any sign of a passage from the older form into the newer?"[46] Falconer answered in the negative. In fact, the dental morphology of these species reveals no sign of a gradual passage nor any variation within the species. Moreover, the example of the mammoth proves exactly the opposite of evolution as described by Darwin: it embodies the stability of a form. "The whole range of the Mammalia, fossil and recent, cannot furnish a species which has had a wider geographical distribution, and at the same time passed through more extreme changes of climatic conditions, than the Mammoth. If [as Darwin thinks] species are so unstable and so susceptible of mutations through such influences, why does that extinct form stand out signally as a monument of stability?"

The mammoth, by its stability and by its great range across space and time, struck Falconer as being a true challenge to the Darwinian theory of descent by means of natural selection. Not that Falconer denied the very possibility of animal species evolving. "I have no faith in the opinion that the Mammoth and other extinct Elephants made their appearance suddenly, in the very shape in which their fossil remains are presented to us. The most rational view seems to be, that they are in some shape the modified descendants of earlier progenitors." But, instead of invoking "natural selection" or "variation from external influences"[47] to explain this evolution, Falconer chose to call for "a more profound and more essential principle," to the action of which "natural selection" would be "but a mere auxiliary." To invoke the physical law that according to him would determine the evolution of forms, Falconer referred to a botanical model, the law of *Phyllotaxis*, which governs the distribution of leaves around the stem of a plant. Falconer did not believe that the influence of Lamarck's internal causes or Darwin's natural selection could explain the transformation of species. Drawing on Goethe and Owen, he preferred to invoke a morphological and logical principle of the successive appearance of forms.

The generation of paleontologists of the 1860s reacted in various way to the presentation of Darwinian theory. Stormy, sometimes violent debates between Huxley and Owen and between Falconer and Huxley[48] represented the collision of two intellectual stances, two generations, perhaps also two different social environments. Whereas Huxley, Darwin's champion, gave talks to workingmen, Owen and Falconer represented the aristocracy or the colonial middle class defending conservative and theological values. But it is probably unwise to push this opposition to extremes. Did the paleontologists reject Darwinism at the philosophical, ideological, and sociopolitical level, or for methodological reasons inherent in the discipline?

In fact, Darwinian theory would continue to be highly controversial and accepted with great reluctance by many paleontologists of the day. The reasons for this may have been inherent in the specificity of paleontology itself. At that time, paleontology did not seem likely to illuminate the processes that Darwin was describing, in part because the very characteristics of its object give little support to proving "the origin of species by means of natural selection." The questions that Falconer put to Darwin raised real difficulties: Where in paleontological material are the traces of variability to be found? How could fossil material prove the processes of the struggle for survival and of "natural selection"?

Beyond the incomplete or unsatisfactory answers given to these questions, one wonders if the Darwinian mode of explanation was actually capable of being the basis of paleontological science at that time. The presence and the succession of fossils were better illuminated from the outside by the evolutionary narratives rather than casting light on the mechanism of Darwinian evolution. In England, the United States, and France, many would reject the accidental, materialistic character of Darwinian evolution in favor of new finalist or vitalist representations of nature and its transformation. It increasingly appeared that fossils told no history whatsoever; they only formed the basis for histories and systems of interpretation that could be built on them.

Of Mammoths and Men

As he passed, he saw an enormous mammoth watching him. It was alone in a stand of young poplars below the riverbank, grazing on green shoots. Naoh had never seen such a big one; it must have been twelve cubits high. A mane as thick as a lion's grew on its neck. The hairy trunk seemed to have a life of its own, half tree, half snake.

The mammoth seemed interested by the sight of the three men; one could hardly imagine it was afraid of them. Naoh shouted:

"The mammoths are strong. The great mammoth is stronger than all the others. It can crush lions and tigers like worms. It can knock over ten aurochs with its chest. . . . Naoh, Nam, and Gaw are the great mammoth's friends."

Flapping its large ears, the mammoth listened to the sounds made by the upright animal. Then it slowly swung its trunk, and trumpeted.

"The mammoth understands," cried Naoh joyfully. "It knows that the Oulhamr recognize its power." . . .

By imperceptible degrees, Naoh drew closer, until he found himself standing before the colossal feet, beneath a trunk that could rip up trees, under tusks as long as an uru's body. It was like being a fieldmouse next to a panther. The animal could destroy him with a single movement. But he stood, trembling with inspired hope, full of creative faith. . . . The trunk brushed Naoh, moved over his body, sniffing him. Holding his breath, Naoh in turn touched the velvety trunk. Then he pulled up some clumps of grass and gave them as a peace offering. His heart swelling with anticipation, he knew he was doing something profound and extraordinary.

Nam and Gaw had seen the mammoth approach their chief, and it showed just how small man was. Then, when the huge trunk lay on him, they muttered:

"Look! Naoh will be crushed. Nam and Gaw will be left alone to face the Kzamms, the beasts, and the waters."

Then they saw Naoh patting the animal, and their souls filled with joy and pride.

"Naoh has made peace with the mammoth!" murmured Nam. "Naoh is the most powerful of men."[1]

This scene from Rosny Aîné's *La Guerre du feu* (translated as *Quest for Fire*) presents the encounter between primitive man and the mammoth, the moment of the pact between the colossal animal and fragile *Homo sapiens* at the earliest moment of our history. The high point in a novel written in 1909, it embodies an archetypal view of prehistory that still resonates with us today.

The mammoth is an intimate part of the imagery of prehistoric man: it is by turns a menacing, terrifying figure; a meek, gentle, and protective being; and a huge beast of prey, the key to man's survival in a hostile environment. The mammoth is an essential witness to humanity's most ancient past, not only because it has served to date man's remains, but also because it was an integral part of his cultures, images, and sustenance. So it is not surprising that the mammoth should have remained connected to our origins as a central image by turns threatening, protective, and supportive.

The idea that man lived in times before written history and that he was the contemporary of "antediluvian" animals is almost commonplace today and taught in every school. But in the first half of the nineteenth century— and especially in France and England—this was the subject of considerable debate involving high philosophical and religious stakes. From the moment that man is no longer seen, as he had been until then, as the ultimate achievement, the crown of Creation, his image becomes significantly debased. Just as the Copernican revolution was for the people of the sixteenth century, this new concept was yet another blow to human pride[2] and the precursor to a major revolution in thought. Starting in the sixteenth century, voyages of discovery and the exploration of unknown lands had stretched our imaginings about humankind. At the beginnings of the eighteenth and nineteenth centuries, Buffon's and Cuvier's great narrative constructions had given an immense time span to the history of the earth and its creatures. But human existence did not really take place within this enlarged time. In both *Epoques de la nature* and *Révolutions du globe,* the stages of this progression and the order of appearance of species on earth matches that of Genesis: a first "epoch" saw the appearance of the lower animals, fishes and crustaceans. Later the land vertebrates appeared, reptiles and gigantic mammals. A final "epoch" witnessed the creation of man on an earth that was at last ready to receive him. For Buffon as for Cuvier—regardless of the differences in their theoretical constructs—the idea remained of an anthropocentric finality of the world, the representation, inherited from religious belief, of man as the ultimate stage of Creation, destined to rule over an earth that had been created for him.

And yet for the philosophers of the Age of Enlightenment, the traditional biblical accounts of Genesis and the Flood were no longer satisfactory answers to the questions about human origins. When and how did man appear? What events led man to become what we are today? These questions were being asked with increasing urgency in the first decades of the nineteenth century. To many, the creation narratives directly inspired by the Bible seemed out-of-date, and new histories were needed to account for the origin of humanity.

Enter the mammoth.

Since the end of the eighteenth century, it was known that the mammoth was a "lost" species and different from today's elephants. It appeared in that gallery of gigantic animals swallowed up by the last catastrophe, which Cuvier had named "antediluvian." To prove the existence of fossil man, it was necessary (contrary to Cuvier's opinion) to show that man—or at least the objects he made with his hands and used—existed at the same time as those great vanished animals. Proving this contemporaneity meant finding man-made tools or human remains in the same stratigraphic layers as the remains of extinct animals. The demonstration would therefore essentially be a geological one.

During the years after Cuvier's death, it appeared more and more probable that man had coexisted in forgotten times with these great extinct "pachyderms." The very phrase "antediluvian man" remained linked to religious belief systems. During the first decades of the nineteenth century, a certain confusion reigned between the term *"diluvium"* (the geological layers that represent the end of the penultimate grand geological epoch) and the religious notion of the Deluge, a confusion Cuvier himself contributed to by characterizing vanished animals as "antediluvian" and by dating the "last and sudden revolution"—the one that, according to him, preceded the appearance of man—at "five or six thousand years" ago. But new representations of man and his evolution, freed from references to religious belief, would soon make that vocabulary obsolete. By the end of the nineteenth century, one would no longer speak of "antediluvian man," but of "fossil man."

Just as the foundation of animal paleontology had been based on rigorous demonstrations and spectacular proofs, the proof of fossil man's existence required facts. In France the search for evidence of man's antiquity was first carried out by marginal seekers unaffiliated with official institutions. They were provincials, and often amateurs.

In 1829 Marcel de Serres, a professor of geology and paleontology at the University of Montpellier, uncovered human tools mixed with the remains of extinct mammals during several excavations in caves in the south of France. In 1827 Paul Tournal, a Narbonne pharmacist who had studied under de Serres at Montpellier, discovered two caves near Bize, north of Narbonne, that contained fragments of pottery and the remains of extinct animals in the same strata. Wrote Tournal: "According to facts that were carefully observed by several people in different locations, man was the contemporary of several animal species that have now disappeared from the surface of the globe."[3] The Belgian doctor Philippe Schmerling had also found, in the Engis Cave near Liège in central Belgium, several human skulls associated with fragments of bone whose points had been sharpened, chipped flint tools, and the bones of fossil animals—mammoths and rhinoceroses. These hominid skulls, some of which were those of children, would not be identified as Neanderthal remains for another hundred years.[4] But in 1835 Schmerling was already convinced of the importance of his discovery.[5]

Though widespread, the belief in the existence of fossil man came under attack by the greatest scientific authorities of the time. In 1832 Lyell claimed that man had been created by "a special and independent attention of the divinity." In the second volume of his *Principles of Geology*, he criticized Tournal's discoveries as having been made in caves, whose layers could have been disturbed. Lyell maintained his position until 1853, in the ninth edition of his book. In France that argument was picked up by Desnoyers, the secretary of the Geological Society of France. Desnoyers saw a stylistic similarity between the tools discovered by Tournal and Schmerling and those found in recent archaeological sites, notably near dolmens. He also felt that the human remains found at those sites were recent ones. If they were found in caves mixed with bones of antediluvian animals, he argued, it was probably because the deposits were disturbed by man, animals, or water.

In the face of criticisms leveled by the scientific authorities of the day, the claims by the pioneers of the 1830s yielded and fell into neglect. They would be rehabilitated only several decades later, when decisive evidence of fossil man's existence was discovered near Abbeville, in Picardy in northern France. This contemporaneity was established not in caves but in open-air sites on the terraces of the lower Somme Valley, with its clearly determined stratigraphy. In 1836 and 1837, a young doctor from Abbeville named Casimir Picard[6] carefully studied the geology of the Somme terraces and the typology of flint tools. This research inspired another local scholar, Jacques Boucher de Perthes, who was the president of the Société d'Emulation d'Abbeville and the town's chief of customs.

Between 1844 and 1859, Boucher de Perthes waged a long battle with the French academic authorities to prove the contemporaneity of stone tools and the remains of extinct animals, a battle that would lead to the recognition of man's antiquity. In his great work published in three volumes in 1847, 1857, and 1864, *Les Antiquités celtiques et antédiluviennes* (Celtic and antediluvian antiquities), Boucher de Perthes distinguished between two successive tool types: "antediluvian" chipped stone artifacts, which were the most primitive and prove humanity's great age; and "Celtic" artifacts of polished stone, which were more recent, coming after the "diluvial." It was precisely by finding the jaw of a mammoth (*Elephas primigenius*) near a chipped flint tool in the oldest layers of the Menchecourt-les-Abbeville terrace in 1842 that Boucher de Perthes for the first time proved that "antediluvian man" existed. "In the month of June 1842 I was brought a white flint [ax] 18 centimeters long by 12 wide, with sharp edges, and of an entirely different shape. The workers told me that it did not come from the bottom but from an intermediate layer. In fact it bore traces of soil from the third layer, which was clayey sand. A few fragments of bone were found next to it in the same clayey matrix. That was . . . part of an elephant jawbone."[7]

The ideas that Boucher de Perthes had so ardently defended would be confirmed by research carried on at the same time on the other side of the English Channel. The attention centered on Brixham Cave in southwest England, which was discovered and explored during the summer of 1858.[8] At Brixham seven chipped axes were found in conjunction with bones of extinct animals. The excavation was conducted by an amateur geologist, William Pengelly, with unrivaled rigor and precision. Though the dig was carried out in a cave, the discoveries seemed undeniable and were followed and supported by the entire English scientific establishment. In 1859 Falconer, Evans, Lyell, and Joseph Prestwich traveled to Abbeville and acknowledged the correctness of Boucher de Perthes's arguments. In the eyes of the entire world, proof that man and mammoth had been contemporaries became an article of faith: fossil man had indeed existed at a time when mammoths, cave bears, and woolly rhinoceroses still ruled the earth.

At that point, the artisan who created those handsome flint bifaces of the Somme Valley was known only by his "works." What did he look like? What was his existence like? To create an image one resorts either to fiction or to building hypotheses based on a few scarce objects. Initially, Boucher de Perthes hesitated between the image of a being fundamentally different from humanity today, a sort of pre-Adamite savage brute, and that of the Adam of the original Garden of Eden. In 1857 he finally abandoned the idea of a break between antediluvian man and modern man. Fossil man was indeed our ancestor. Like us, he was intelligent, sensitive, and artistic. He knew how to shape tools and hunt, and he was capable of appreciating that which is useful and beautiful.

Having discovered signs of fossil man, Boucher de Perthes hoped to find the remains of the man himself. The Moulin-Quignon jaw exhumed in a "lower layer of the *diluvium*" (it was actually a modern jaw fraudulently buried in the ancient layers) lent reality to an almost mythical image of a primitive man who resembled his modern fellow. The "antediluvian man" invented by Boucher de Perthes was a singular mix of conventional ideas and imagination, of mythology and rational constructs.[9] But prehistory, the science of prehistoric human cultures, soon spread far beyond the limits its founder had assigned it. Teaching and research centers, journals, and museums quickly began to appear in France, England, and other European countries. The touchstone of this new discipline, Lyell's *The Geological Evidences of the Antiquity of Man* (1863),[10] summarized the research of an entire generation. And human paleontology—the study of skeletal remains of fossil man—soon emerged as a new scientific domain to rival animal paleontology.

What was involved in the debates of those years was more than the creation of new knowledge and new disciplines. It was the appearance of a new representation of the history of the world in which man was no longer the crown of Creation and destined to rule over it, but a living being whose history is intimately linked to those of other animals. While some still tried to set the quest for fossil man within a framework that more or less respected the Scriptures, others were beginning to look at the question of human origin in new ways. Perhaps man, who resembles the great apes so closely that Linnaeus classified him in the primate order, did not originate in a single, creative event ex nihilo, but in a slow evolution rooted in his animal nature. That was Lamarck's theory in 1809,[11] and it would become a Darwinian concept by midcentury. But in the *Origin of Species* in 1859, Darwin says very little about man's origins. One would have to await the works of his disciple Thomas Henry Huxley in 1863, those of the German biologist Ernst Haeckel in 1874, and the publication by Darwin himself of *The Descent of Man* in 1871[12] for the theory to be set forth clearly: *Man is descended from the Apes.*

In the last decades of the nineteenth century, this formulation was both sensational and scandalous. In Victorian England, the idea was shocking enough to provoke in 1862 a violent public argument between the two most eminent English paleontologists of the day, Richard Owen and Thomas Henry Huxley.[13]

To say that man was the contemporary of mammoths gives only a relative sense of his antiquity, and arguments about fossil man's absolute age very quickly arose. The question was heavy with implications not only for science, but also for religion. While some tried to preserve the traditional framework of the Bible's six thousand years, others relied on calculations based on geological observation to give human history a far greater duration. Lyell, who believed that slow and uniform causality was at work in the history of the earth, estimated the age of man at 100,000 years. But Joseph Prestwich, with his alternation of slow causes and more brutal events (such as deluges), suggested an age of 20,000 years. These arguments would not be finally settled until nearly a century later with the development of radiocarbon dating in the 1950s—but humanity had forever been enriched by a new dimension, that of time.

During the second half of the nineteenth century, prehistorians focused on establishing a chronology of prehistoric times. What were the stages of prehistoric man's development? What about his physical evolution, his intelligence, his cultures? The French paleontologist Edouard Lartet (1801–1871) suggested that one could distinguish the stages in the history of early man by observing the successive changes of the "characteristic fauna of this long period known as Quaternary or diluvial."[14] According to Lartet, the cave bear was proof of a first epoch, earlier even than the "diluvial terrain of Abbeville"; its extinction would therefore have preceded that of all other Quaternary species. The mammoth represented a second epoch, having in the Tertiary come from Asia (as it was then believed), where it lived; the mammoth (*Elephas primigenius*) and woolly rhinoceros (*Rhinoceros tichorhinus*) lived in Europe when Neanderthal man appeared. The next period was that of the reindeer, and then, in historic times, the age of the aurochs.

But it was soon noted that defining the "ages of man" by the succession of contemporary fauna did not fit the facts. The remains of great bears and mammoths, which represented the first two epochs, occurred fairly often during the other stages of the Pleistocene. To complicate things further, fossil human remains were found associated "with another species of elephants (*Elephas antiquus*) whose extinction was supposed to have been earlier."[15] Moreover, the presence or absence of fossil elephant remains in the various sites depended on such parameters as "the relief of the soil, its altitude, nature, and distance from the sea." So a chronology established in function of variability according to animal remains would not allow one precisely to distinguish the successive periods of human prehistory. Man and mammoth had clearly coexisted for a very long time, perhaps for most of the Paleolithic in Europe, and not just for a brief period between what Lartet had called the ages of the great bear and the reindeer.

Lartet's paleontological chronology therefore had to be abandoned. Instead, fossil man's history was traced indirectly through his artifacts, which revealed the development of his intelligence. Among the artifacts Boucher de Perthes had collected in the Somme terraces, he had distinguished between Celtic tools, the most recent ones, and antediluvian, the most ancient. In 1865 the English anthropologist John Lubbock[16] came up with the terms "Paleolithic" to designate the epoch of chipped stone, which was the oldest ("during which man shared Europe with the mammoth, the cave bear, the woolly rhinoceros, and other vanished animals"), and "Neolithic" for the more recent polished stone epoch. But Lubbock's classification distinguished only two stages of human prehistory. The human fossil remains, the tools of worked stone, and the artistic evidence that were beginning to accumulate suggested that this history may have experienced successive "ages," which could be more accurately identified.

In 1865 Gabriel de Mortillet, who classified Boucher de Perthes's collections for their display in the museum of Saint-Germain-en-Laye, suggested a new nomenclature of the ages of prehistoric humanity based on the function of the tools that characterized each period.[17] Mortillet thought of himself as a materialist and a Darwinian (though he probably only knew about Darwin through Albert Gaudry's work) and conceived of human and cultural evolution as a progressive transformation of prehistoric cultures and human types, with the following successions: first, the Saint-Acheul or Acheulean period, the most ancient, characterized by bifaces; then the Moustiers (a Dordogne Valley site) or Moustierian epoch, characterized by chipped stone tools; the Solutré or Solutrean epoch (from the site near Macon), characterized by finely worked "laurel-leaf" tools; and finally the Madeleine or Magdalenian epoch, named for the Madeleine site in Dordogne, for the end of the Paleolithic, characterized by more finely worked flaked tools and arrowheads. This method had the advantage of linking the successive types of artifacts to the various prehistoric human types that were known at the time: to the Acheulean corresponded the "Moulin-Quignon type" and later the jawbone discovered by Mauer in 1907; to the Moustierian, Neanderthal man; and for the last two, the Solutrean and Magdalenian periods, *Homo sapiens sapiens*, represented by the famous Cro-Magnon man discovered in the Vézère Valley in 1868.

Thus, prehistoric studies of the second half of the nineteenth century produced a fairly simple chronological framework of man's prehistoric genesis. An attempt was also made to find correspondences between man's evolutionary and cultural history and the environmental, geological, and climatic phases that were then known. (At the time, geologists recognized two main successive climatic phases: a warm phase followed by a cold one ending in a slow warming trend leading to modern times.)

In the first decades of the twentieth century, however, this carefully calibrated system would fall apart. The vision of human evolution was changed by various discoveries concerning hominid remains, tools, art, and the environment, as well as the broadening of research to a wider geographical area. Linear scenarios gave way to observations that stressed the existence of diversity, superpositions, and the coexistence of several types of hominids and several types of cultures.

In this burgeoning of species, what was the "true" origin of man? With the discovery of australopithecines in South and East Africa in 1925, it seemed more and more certain that the cradle of humanity was neither in Asia nor Europe, but in Africa. And man had aged beyond the 100,000 years Lyell gave him at the end of the previous century. That figure—which appears boldly in the epigraph of *Quest for Fire*—grew to a million years and more. It is remarkable that establishing the ancient chronology of the first African hominids began with African fauna and in particular the fossil proboscidean species.[18] In this complex and now immensely lengthened history, the woolly mammoth represents only a recent period of human history (middle to late Paleolithic) characterized in Europe by an alternation of glacial and warming phases. Having appeared 400,000 years ago, the mammoth lived in the European region along with the last *Homo erectus*, the first Neanderthals, and the first *sapiens*. Like other cold-weather species—the cave bear, reindeer, and hyena—it disappeared when the great glaciers melted.

The woolly mammoth has come to symbolize the whole of prehistory because it is associated with the last phases of that history, which were both the coldest and the most human. But the picture is far more complex than it was in the heroic days of the search for "antediluvian man." It now appears that during the course of its history, the mammoth coexisted with different species of hominids. And during the course of human evolution, several different species of fossil elephants existed that have now disappeared.

The investigation of the existence and the meaning of prehistoric art marked an essential stage in shaping an image of fossil man. Was primitive man capable of appreciating beauty? Did he produce images only as an imitation of reality, or could his art also manifest religious thought? Was its meaning related to his mode of subsistence? These questions, and the answers given to them, reveal the changing representations of prehistoric man.

After 1860 Boucher de Perthes's "figure stones" were forgotten in favor

The Madeleine (Dordogne) mammoth, inscribed on mammoth ivory. This engraving, found by Lartet and Christy in 1864 in a late Paleolithic site, was virtually conclusive proof of fossil man's existence. (© MNHN, Paléontologie, D. Serrette photo.)

of authentic discoveries of objects made of bone, ivory, and reindeer antler, carved or inscribed, from the late Paleolithic. The most famous discovery of the time was made in 1864, when Edouard Lartet found a sheet of mammoth ivory with a picture of an actual mammoth scratched into it. For anyone who still required proof, this was a discovery that brought together all the evidence proving that man and the great vanished "pachyderms" had been contemporaries. Paleolithic man had not only lived alongside the mammoth; he had used its ivory to sculpt and inscribe his works of art. On part of one of its tusks, he had even scratched the portrait of the animal itself, looking much like the frozen mammoth that had been extracted practically whole from the Siberian permafrost! This discovery proved the humanity of these "primitive" people, who henceforth would be recognized of capable of producing true thought and art.

Falconer and Christy were visiting Lartet in Dordogne at the Madeleine excavation in May 1864 when his workers brought him "five broken pieces of a fairly thin sheet of ivory" that they had dug up. Lartet described the find:

> Having joined the pieces together, according to lines of junction marked out by the minute intricacies of fracture, I showed Dr. Falconer the numerous characteristic, though shallow, engraved lines, which seemed to me to indicate some animal forms. The practiced eye of the celebrated palaeontologist, who has so well studied the Proboscidians, at once recognized the head of an Elephant, and he soon pointed out other parts of

the body, and particularly, in the region of the neck, a bundle of descending lines which recall the long shaggy hair characteristic of the Mammoth, an Elephant of the Glacial Period.[19]

Other ivory engravings and sculptures were soon discovered. Between 1894 and 1897, Edouard Piette explored "the most ancient reindeer age strata"[20] (it corresponds to the Gravettian) at Brassempouy in the Landes, in southwest France, and found huge deposits of ivory in its lower layers. During this period of Western prehistory, ivory was starting to replace bone, not only for such tools as needles, punches, and arrowheads, but also as bases for carved or sculpted art objects, inscribed sheets, and statuettes—"human figurines of remarkable workmanship and strange realism." To this period, which is characterized by the abundance of ivory used for tools and art, Piette gave the name of the "elephantine" or Eburnean period. Wrote Piette:

> The Eburnean epoch began at the time when the great Mouste-
> rian glaciers began to melt. The temperature was becoming
> milder. . . . Still quite ignorant, [man] had only chips of flint
> for tools. With these fragile and imperfect implements he
> shaped many utensils out of bone and carved mammoth ivory.
> He sculpted statuettes to represent the beings around him. On
> sheets of ivory, which were too thin to be sculpted in the
> round, he carved reliefs in which appear not only animate
> beings, but also decorations produced by his imagination. Art
> especially flourished in the foothills of the Pyrenees, near the
> [Atlantic] ocean and the Mediterranean.[21]

From these layers Piette excavated a large number of ivory figurines mixed with the bones of mammoths and woolly rhinoceroses. Among them in 1894 he found the famous Brassempouy "Dame à la capuche," a tiny sculpted face with fine, delicate features, topped by a headdress decorated with a hairnet. It is today one of the prizes of the Musée des Antiquités Nationales at Saint-Germain-en-Laye. Piette also dug up other sculptures representing the female body, with slender or opulent shapes. "Enamored of human forms, the inhabitants of these regions excelled in representing women," he wrote. "It is as if love incited the first sculptor to bring ivory to life along the traits of his beloved. Some of these statuettes have no more artistic value than dolls, but some are admirably modeled. These people were true artists."

Given the abundance of animal and human figures, the question then became one of their interpretation. Was primitive man driven to represent his fellows out of an awakening of instinct, sentiment, desire, or need? Could these artistic expressions be linked to prehistoric religion? Or did he practice "art for art's sake," as one did in 1880?

At the start of the twentieth century, other examples of Paleolithic art were being discovered in western Europe, though the painted or carved cave art was initially met with skepticism. The Altamira Cave, which was discovered in 1875, would not be recognized as authentic until 1902. From then on the discoveries of painted caves became more numerous, especially in southwestern France. Animal shapes were found drawn or painted on the walls of the caves at La Mouthe, Font-de-Gaume, Les Combarelles, Bernifal, and Pech-Merle. Mammoths are present but relatively rare in these representations of Paleolithic art in western Europe. One intriguing exception is the Rouffignac Cave in Dordogne, "the cave with a hundred mammoths" drawn or scratched on its walls. The rarity of mammoths in western Paleolithic cave art was part of the argument over the authenticity of these cave drawings that erupted when they were discovered in 1956. If they are authentic (and a number of prehistorians today say they are), the Rouffignac mammoths are without any doubt the most numerous, the most expressive, and the most moving.[22] Since the beginning of the century, many attempts have been made to interpret this animal art as expressing religious values or as part of hunting rituals. To André Leroi-Gourhan, the "iconographic system of western Paleolithic art" is structured according to a binary split that he thinks embodies sexual dualism. The mammoth is the "third animal" that is occasionally associated with the bison-horse couple arrangement of animal figures on cave walls.[23] The mammoth's rarity contrasts with the abundance of its images in the prehistoric art of eastern Europe.

In 1920 Czech and Russian discoveries again pushed the mammoth to the front of the stage of human prehistory, by revealing the existence of a veritable "age of mammoths" in central and eastern Europe, contemporaneous with the "age of reindeer" in western Europe and of the epoch when *Homo sapiens sapiens* appeared, thirty thousand years ago.[24] In those parts of Europe that were covered by glaciers in the middle and late Paleolithic, certain sites were known to contain extraordinary quantities of gigantic bones. In Moravia, Poland, Ukraine, and Siberia, the names of certain localities echo the fact that bones have always been found there: Mamutovo (Poland), Kostienki (Russia), and the Předmost bone fields in Moravia. The Kostienki site was noticed as early as the eighteenth century by the German naturalist Gmelin during a journey of exploration in Siberia and Russia. He saw huge piles of elephant bones at a site on the right bank of the Don River not far from the town of Voronezh. In 1880 *Anthropological Travels in Central and Eastern Russia* by the naturalist Ivan S. Poliakov was published in St. Petersburg. In the second chapter of the book, Poliakov described his visit and excavations at the Kostienki site near the village of Karatcharov in 1878.[25] The following year he discovered a prehistoric dwelling site in the main street of the village, where mammoth bones were accompanied by characteristic chipped Paleolithic flints.

> Like the tools, the bones were not scattered randomly. In some places none were found, in others they were gathered together. In certain places they were so mixed up to the point that if bones were found there would also be tools and other traces of human presence—and to the contrary, the fewer tools there were, the fewer the bones. This was so striking that my workers, who were generally skeptical about my projects and what I said, were quickly convinced. While I was as usual looking for carved stone tools in the clay dug up from the bottom of the pit, I would often hear one or another of my workers shout, "Baryne, there are bones here!" Whoever was shouting knew that he would soon find flint tools. I then had to climb down into the pit and search with my own hands while carefully looking over the remains already gathered by the workers. Later, I would hear "Baryne, there are stones here!" which meant that bones would soon be found, and I would have to again climb down into the pit.[26]

As with the sites of western Europe, these collections of mammoth bones and tools proved the existence of fossil man in these areas during prehistoric times. Poliakov's research and excavation methods remind us of the epic period of early Russian prehistoric archaeology, inspired by the methods and ideas of French prehistory (Poliakov even speaks of "Acheulean" . . .).

A few decades later, between 1920 and 1930, the "Marxisation" of Soviet archaeology would impose a new vision of the progress of human culture. In 1919, in the middle of the civil war, Lenin signed the decree creating the Academy of the History of Material Cultures. Nikolay Y. Marr, a linguist and archaeologist from the Caucasus, was named its director. Drawing on the writings of Lewis Morgan and Friedrich Engels about the transformation of primitive societies, Marr worked out a theory of the "stages of development" of primitive societies and applied it to archaeological research. "Marrism" interprets the succession of "cultures" founded on topological criteria—the Aurignacian, the Solutrean, the Magdalenian—which French prehistorians had described, as a succession of socioeconomic stages, reflecting the progress of the means of production and the transformation of the forms of work in this early epoch of human history.

This conception of evolutionary stages of primitive societies meant that the practices of archaeological research had to be rethought. Soviet archaeologists developed new excavation methods associated with these views of man and his history. They made it possible to uncover successive layers of habitation by the technique of horizontal excavation.[27] Examining successive horizontal layers made it possible to account for the stages of evolution of prehistoric societies. On this theoretical and practical basis, P. P. Efimenko and S. N. Zamiatnin in 1923 started excavating the Kostienki I site—a veritable laboratory of Soviet and Russian prehistoric archaeology. Their aim was to transform prehistoric archaeology into a historic science, in accordance with the Marxist-Leninist principles. The purpose of the dig was to "reveal a organized complex that had economic significance."[28]

From that period on, the method of horizontal excavations made it possible to uncover organized groupings. One of them appeared as "an oval concentration of about 2,000 large mammoth bones, whole or broken, some of them vertical, covering an area 25 by 20 feet and 1.5 feet deep, arranged according to the type of bone."[29] These oval groupings suggested habitation sites, whose architecture and layout could be reconstructed. All the huts' entrances were oriented roughly toward the fire pit at the center of the dwelling. The oval field was divided lengthwise by a row of fire pits and scattered with pits and holes of all dimensions and purposes. The largest and deepest pits were located on the edge of the oval field. Four of them could have served as underground dwellings, while twelve smaller, rounder ones, which were filled with mammoth bones, could have been garbage pits."[30]

The tusks and flat bones of mammoths were used to build roofs. On the surfaces of the dwellings' floors, many figurines of ivory or carved limestone were found, primarily female figures and stylized mammoth shapes. The Gravettian female statuettes have the same opulent curves as those found in western Europe. One mammoth ivory carving, whose facial traits are well rendered, is particularly expressive. The figurines representing mammoths are fairly conventional and generally carved in marl.[31]

Piotr Petrovitch Efimenko, in a celebrated 1931 article entitled "Significance of the Woman in the Aurignacian Epoch,"[32] picked up the idea—in conformity with Marrist doctrine—that matriarchy characterized the first socioeconomic stage of primitive societies. The feminine images incarnate both the social and economic role of the woman in these cultures, but also her spiritual role in an ideology built around women-mothers within a matriarchal social organization. Zoia Abramova explains: "The image of the woman as captured by these statuettes shows the important role played by woman-mother in the community of the late Paleolithic. She represents both the woman as mistress of the household and its vital fire, as well as the ancestral woman connected to the idea of a woman whose magical powers can ensure the success of one of the main subsistence activities, hunting."[33]

Relying once again on Efimenko's work, Abramova remarked in 1979 that the female Gravettian figurines at the Russian sites were often associated with very small and conventional representations of mammoths. Moreover, the female statues were always found in places that contain large quantities of mammoth bones. Women and mammoths are equally associated in sculpture and in plastic representation. The images of the mammoth are as frequent as they are because the subsistence of settled communities depended entirely on that animal. The conventional animal figurines that remain symbolize this "need to hunt, the economic basis of prehistoric man." "Beyond the economic role of the mammoth in daily life, there is evidence to suggest the considerable importance of the mammoth in the spiritual life of prehistoric man. For example, the use of [mammoth] shoulder blades in funeral rites. . . . At Brno, a [human] skeleton was found covered by a mammoth tusk and shoulder blade."[34]

On the Russian plains, these characteristic traits—hunting economy, sedentary existence, and matriarchy—are linked to a characteristic economic stage of the Aurignacian period. Woman represented the social structure and the societies' economic values, and the mammoth played a major economic and symbolic role.

In 1950 the stages theory of Marxist archaeology began to be criticized by its own theoreticians. A major debate took place between the archaeologists Efimenko and Rogatchev, and Stalin published a virulent article in *Pravda* attacking Marrs's linguistic theories, which he denounced as "perversions of Marxism."[35] By the time Stalin died in 1953, the theoretical framework of stages archaeology had already been largely abandoned. Moreover, it was recognized that the typological classification established for French archaeology had no meaning for the cultures of eastern Europe. But the techniques of horizontal excavations were maintained and applied to an approach to prehistoric cultures that was now *ethnographic*.[36] The Kostienki I site, which had been studied and excavated by a team of archaeologists led by Efimenko,[37] continues to play the role of a laboratory of Soviet and then of Russian prehistoric archaeology up to the present.[38]

Around the same period, the sites of Mežerich in Ukraine, Dolni Vestonice in Czechoslovakia, and Předmost in Moravia were explored.[39] In 1945 the Czech historian Karl Absolon did important research at Dolni Vestonice. What had been initially believed to be garbage heaps or the effects of a natural cataclysm were identified a few years later as the remains of dwellings. The

A Paleolithic dwelling made of mammoth bones at Mežerich, Ukraine, between 14,000 and 18,000 years ago. (N. D. Praslov photo.)

interpretation of these accumulations of bones as habitation sites was also made possible in those sites by the development of horizontal methods of digs. In 1950 new digs led by the Czech archaeologist Klima uncovered dwelling structures. Huts, whose walls were probably covered with animal hides, were built on bases of mammoth skulls and jawbones. In these frozen wastes, where wood was rare or entirely absent, mammoth remains were used for everything. Even their bones served as fuel.

The Pavlovian camps in Moravia consist of huts whose perimeter is made of the lower jawbones of mammoths arranged in chevrons. These huts contain decorative objects of all sorts, ornaments, and even furniture, made of mammoth bone or ivory. At Mežerich in Ukraine, a skull decorated with ochre may have served as a stool or drum; two mammoth skulls above a fire pit could have held a spit for grilling meat. A musical instrument—a kind of xylophone—was made from mammoth shoulder blades. In a Moravian Paleolithic sepulture, a mammoth tusk and shoulder blade were even used to cover the corpses of two young girls. The mammoth was an "all-purpose" resource. Its flesh was used for food; the hair and skin for bedding and clothing; the ivory for ornamentation, as pendants and necklaces, for art, decoration, and building; and the bones in construction and for fuel. One can truly speak of "civilizations of the mammoth" in these regions, since the mammoth was the basis of life and of material survival itself, and these civilizations disappeared at the same time as the mammoth, around 12,000 B.P.

These discoveries gave a new dimension to the history of the relationship between man and mammoth and to their images. In central Europe a sedentary prehistoric civilization had existed from the oldest period of the late Paleolithic. This in turn brought into question the conventional view of prehistoric man in the Aurignacian or Gravettian period as a nomadic hunter-gatherer. At that early period, in the beginning of the late Paleolithic, representatives of our species, *Homo sapiens*, were already living in open-air habitations and using baked clay more than twenty-five thousand years ago—some fifteen thousand years before its appearance in Neolithic sites. New scenarios evolved to show how mammoths were hunted, in which weakness defeated power and cunning matched these colossi and their strength and supposed intelligence and memory (extrapolating from those of elephants)—all of which made mammoth hunting an emblematic image in the representation of prehistoric man.

Things get complicated when one starts asking specific questions, however. What weapons did these people use? What techniques did they use to hunt these enormous animals? "Short-range attacks do not seem out of the question," wrote Kurt Lindner in a 1937 book that was long considered authoritative. "Like the elephant, the mammoth had small eyes and probably did not see well. Hidden in tall grass, an experienced hunter would be able to creep quite close to the game."[40] Likewise the theory of weighted traps also was in vogue for a period. Lindner thought he saw evidence for them in certain cave paintings and engravings of the upper Paleolithic. But the use of weighted traps in mammoth hunting remains controversial. How much weight would be needed to knock out a mammoth, given the volume of its skull and the thickness of its bones? Other scenarios suggested that mammoths were driven off the edge of cliffs like the horses from the heights of the Solutré cliff. A painting illustrating this improbable fiction has long hung on the walls of the Geological Institute of St. Petersburg.

One can also imagine bands of humans trailing mammoth herds along their great seasonal migration routes, collaborating to cut off and exhaust individual animals and bringing them down by wounding them at their weak points—eyes and neck—perhaps with poisoned arrows. Or maybe the mammoths were driven toward pit traps, where they were riddled with projectiles and killed. Some of Zdenek Burian's paintings have captured these images and contributed to their persistence. But the theory of pit traps is conceivable only for small, young animals, and it is hardly plausible in the regions of central Europe where "the frozen soil of the steppe can only be dug with

The artist and his work: *The Creator of the Předmost Venus,* by Zdenek Burian. (Oil on canvas, 1958.) (Moravské Zemské Muzeum–Anthropos, Brno.)

difficulty."[41] That man hunted mammoth is supported by quite a bit of archaeological evidence, including arrowheads stuck in the bones of young mammoths. But this prehistoric hunting epic does not take into account the fact that the enormous quantities of bones in the eastern Europe sites could have accumulated over a period of thousands of years. It is possible that only a small quantity of individual animals were hunted in any given year and that the mass slaughter that is often described never took place.

The American archaeologist Lewis Binford challenged one of our strongest images of prehistory when he provocatively suggested that Paleolithic man was actually a scavenger who fed on the carcasses of dead animals.[42] But Binford's model concerns the oldest periods of human prehistory. For the late Paleolithic, the myths of the mammoth hunt as they were constructed in the nineteenth and the first half of the twentieth century have remained current.

The last mammoths migrated beyond the areas inhabited by man, to the islands north of Siberia and to the Alaska mainland, where they finally died out—and where their bones now litter the earth. The history of the relations between man and mammoth comes to a close at the end of glacial times and the end of the epic of the great Paleolithic hunters, whose memory is strikingly preserved on the walls of the caves of Altamira, Lascaux, Rouffignac,

The Mammoth Hunt, by Zdenek Burian. (Moravské Zemské Muzeum–Anthropos, Brno.)

Chauvet, and Font-de-Gaume. The disappearance of the mammoth marks the end of an epic period in human existence, when man lived in the cold, glacier-covered regions of Europe and North America. The mammoth remains the symbol of these periods, which for us merge into the early known history of humanity. But the image of the proximity of man and mammoth is also deeply rooted in the search for the prehistoric origin of man and his cultures in western and central Europe, Asia, and North America.

PART FOUR
SCENARIOS

The Mammoth in the Trees

Elephants evolve. Living species have been subject to transformation over an immense period of history. With the Darwinian theory of evolution, the arrangement of beings in nature today and the succession of fossils in the layers of the earth are given meaning within a genealogical history of life. "Propinquity of descent, the only known cause of the similarity of organic beings, is the bond . . . which is partially revealed to us by our classifications," wrote Darwin.[1] If the diversity of the living world today is the result of a continuous history, the classification of living and fossil species must account for the genealogical relationships that species have with one another, and for the history of their transformations and diversification over geological time. The image of a ladder has been replaced by that of a tree with spreading branches. This image can be read vertically, as the history of a group that evolves and diversifies across geologic ages, and horizontally, as representing the coexistence of living species at a precise moment in that history. In this way, Darwin resurrected the old biblical "tree of life" image to describe the evolution of species "by means of natural selection."

> The affinities of all the beings of the same class have sometimes been represented by a great tree. I believe this simile largely speaks the truth. The green and budding twigs may represent existing species; and those produced during each former year may represent the long succession of extinct species. At each period of growth all the growing twigs have tried to branch out on all sides, and to overtop and kill the surrounding twigs and branches, in the same manner as species and groups of species have tried to overmaster other species in the great battle for life. The limbs divided into great branches, and these into

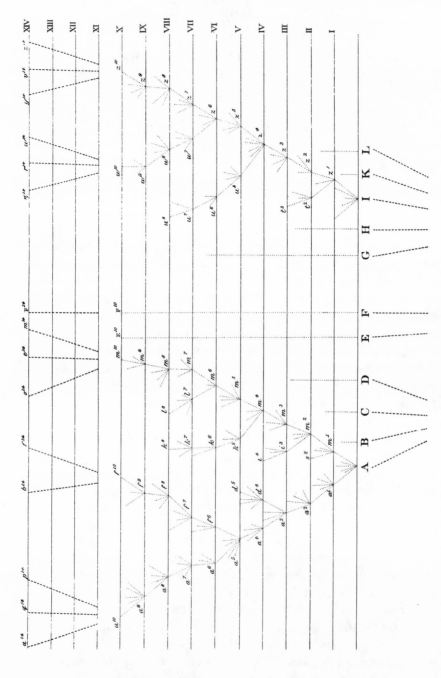

The evolutionary tree, a theoretical schema drawn by Darwin in *On the Origin of Species* (1859), chapter 4.

lesser and lesser branches, were themselves once, when the tree was small, budding twigs; and this connection of the former and present buds by ramifying branches may well represent the classification of all extinct and living species in groups subordinate to groups.[2]

Darwin presented the image of this treelike schema of descent in chapter 4 of the *Origin of Species*. But it was only an abstract model, and he invited paleontologists to go in search of fossil remains that would reconstitute the evolutionary history of species. Starting in 1860, such paleontologists as Thomas Henry Huxley in England, Othniel Marsh in the United States, Albert Gaudry in France, and Vladimir von Kovalevsky in Russia actively started hunting for the fossil remains of these "missing links" and tried to organize them into a genealogy. Despite the gaps in the fossil record, the paleontological history of species was beginning to be reconstructed.

Here again, the mammoth played a part. Because of its profusion and complexity, the order of Proboscidea is a prime object for taxonomists. Cuvier had defined a single species of "fossil elephants" and said that the extinct genus of mastodons was represented by "one or two species." But starting in the nineteenth century, the number of fossil proboscidean species increased dramatically. Each of the "phylogenetic trees" that has been attempted suggests a different genealogical schema according to the characters studied, the forms known, the supposed filiations, and the evolutionary philosophy behind the schema. Each of these classifications places the various species in relationships with hypothetical common ancestors and within geological epochs and precise locations. Each takes into account the discovery of new fossils. And each, relying on often divergent representations of evolutionary processes, tells a different story.

The French paleontologist Albert Gaudry was one of the first to build genealogical "trees" to represent the phylogenetic history of fossil forms, taking their stratigraphic position into account. As early as 1866, he illustrated his *Animaux fossiles et géologie de l'Attique* (Fossil animals and geology of Attica)[3] with pictures representing the phylogenies of several families of land vertebrates. His genealogical tree of the Proboscidea—almost certainly the first of its kind—is a snapshot of the paleontological knowledge of the day. It also expressed his philosophy of the evolution of the living world.

"Paleontology today has taken on immense proportions. It has become the history of the development of organic nature. To properly grasp this history, it would be necessary to build a long gallery where we would follow the development of life, starting from the day of its first manifestation on the globe

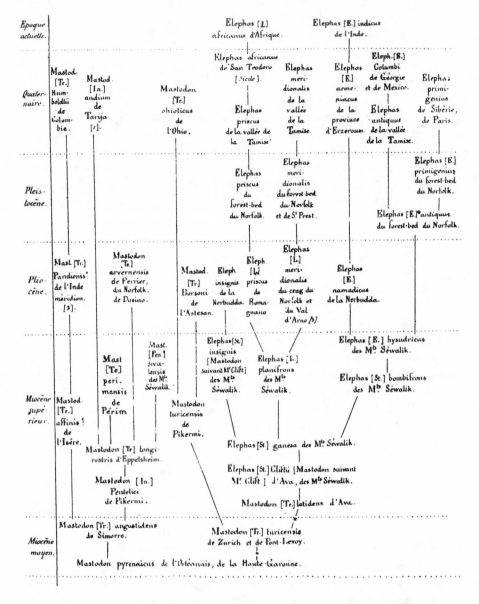

The evolutionary tree of the Proboscidea, according to Albert Gaudry, *Considérations générales sur les animaux fossiles de Pikermi* (General considerations on the Pikermi fossil animals) (1866), p. 38.

to that in which human intelligence shines forth," wrote Gaudry in his introduction to a text entitled "Sur l'éléphant de Durfort,"[4] his contribution to a volume published in 1893 to celebrate the centennial of the founding of the Muséum National d'Histoire Naturelle in Paris. At the close of the nineteenth century, Gaudry was the founder of the great gallery exhibiting the Muséum's paleontological collections. In this enterprise he was no doubt inspired by his visit to the Hunterian Museum of Glasgow, but also by the "strange" and powerful "impression" that he got from his visits to the paleontological museums founded by the great American paleontologists at Pittsburgh, New Haven, and New York, which for some years had held the gigantic remains of mammoths and dinosaurs discovered in the American West.

The paleontology gallery was built during the last years of the century according to a plan by the architect Duterte, who had been in charge of public works for the 1889 Universal Exposition. Inaugurated on July 21, 1898, the gallery was a huge public success. Eleven thousand visitors crowded in on the first Sunday, ten thousand the next, as curious Parisians discovered in the gallery the marvels of the fossil world. After climbing the wide staircase to the vast hall, they encountered a troop of fossil animals with brownish bones, strange shapes, and sometimes barbarous-sounding names: mastodon, paleotherium, *Triceratops*. Dominating this array of skeletal monsters, with its gigantic feet and enormous size, was "the Durfort elephant," which had been dug up in the Gard region of southern France between 1869 and 1873.

The arrangement of the fossils in the Muséum's paleontology gallery—from the most ancient to the most recent—illustrated "the history of developments of organic nature" according to Gaudry's evolutionary concepts. Unlike many French naturalists, who remained faithful to Cuvier's fixism, Gaudry believed in the evolution of the living world. Beginning with his work in the middle of the century, Gaudry proclaimed his admiration for Darwin, even though it led to his being isolated both at the Sorbonne and the Muséum.[5]

Gaudry ended the opening lecture of his course in paleontology at the Muséum in 1873 by proclaiming:

> Along with the differences presented by species of consecutive epochs, many points of resemblance seem to reveal their filiations. Paleontology is beginning to glimpse great chains of organized beings, which stretch across geologic ages to connect the creatures of ancient days with those of today. Many of the links of these chains have been lost to us forever, but some have been discovered or rediscovered. We will energetically seek

them together, because the history of nature is dominated by a plan and I am confident that the study of paleontological links will help us understand that plan, which is still mysterious.[6]

When Gaudry in 1863 declared his belief in evolution, he expressed it as a necessary progression of beings, rather than invoking the Darwinian principle of "the origin of species by means of natural selection." With Gaudry, the search for intermediate links took on a spiritualist coloring, marked by the traditional representation of the hierarchical "chain" of beings, and an idyllic vision of the harmony of the world, where everything is linked in a perfect accord. He supported Darwin but remained attached to a metaphysical vision of evolution determined by a divine "plan," which he conceived as a necessary progression culminating in man.

The possibility of proving the existence of "intermediate" species connecting genera that until then had remained distinct was born of a better knowledge of fossil species and an increase in the number of excavations. In 1867 Gaudry published the results of the digs he had carried out in Greece starting in 1855 on Miocene fossils in Pikermi.[7] In that fauna he encountered new types, which he thought were true intermediaries between genera or even families of mammals. A second book, *Animaux fossiles du mont Lubéron* (1878) (Fossil animals of Mount Lubéron), is devoted to a site in southeast France contemporaneous with Pikermi, and it stresses the existence, within fossils species, of a variability that struck Gaudry as significant enough to justify the distinction of races. He drew on this fieldwork for his great works of synthesis. In his *Les Enchaînements du monde animal* (Linkages of the animal world) (1878–90), Gaudry tried to apply a transformist method to the animals of the Primary, Secondary, and Tertiary periods.

Mammoths and elephants occupy a special place in his "evolutionary philosophy," which he outlined in an 1896 work entitled *Essai de paléontologie philosophique* (An essay on philosophical paleontology). In revisiting the traditional division of Proboscidea between mastodons and elephants, Gaudry showed that "intermediate" species exist that in the course of their history have truly played a transitional role between these two genera. "In several parts of Europe teeth are found that are certainly those of elephants, but which suggest those of mastodons because of the small number of their bumps, their width, and the thickness of their enamel plates. They are characteristic of the late Pliocene. We see other teeth where the mastodon-like character disappears, the bumps become more numerous and wider, and the enamel becomes thinner."[8] In this way an "intermediate form" discovered at Pikermi, *Mastodon turicensis*, could fill the hiatus that separates mastodons and elephants, and extend the uninterrupted linkage of forms in the history of the Proboscidea family.

What about "that gigantic creature," the Durfort elephant? What place should it be assigned in the evolutionary history of the proboscideans? Gaudry worked to place it in relation to the other known species of fossil elephants—*Elephas antiquus, Elephas meridionalis,* and *Elephas primigenius.* From anatomical study, he concluded that it belonged to the species *Elephas meridionalis,* but to a "modified race" of the species. "It seems to me . . . that the tusks are more curved and the bones of the feet less thick than in the oldest *Elephas meridionalis.* In this, the Durfort animal marks yet another tendency toward the Quaternary elephants." The Durfort elephant was a "more evolved" variety, he thought, pointing toward future species. The study and reconstitution of the fauna and flora contemporary with this elephant confirmed its intermediate character: it belonged to the epoch "that marks the transition from the Tertiary to the Quaternary."

During this period, nature displayed a "majestic and pacific spectacle," and the animal itself appeared "as the most imposing mammal skeleton encountered until then." At the end of the Tertiary, elephants were the kings of this fauna, with their majestic appearance, due not only to their size, but because they stood on "legs as straight as columns," their heads raised to the sky. Thanks to their trunks, "elephants do not need to bow down toward the earth to seek their nourishment, as other ungulates must." Gaudry concluded that in the animal world, the Proboscidea had a posture that seemed to prefigure man's upright stance.

During the course of geological ages, each epoch witnessed a kind of perfection, of grandeur. The next epoch would culminate in the arrival of *Homo sapiens.* Thus the elephants at the end of the Pliocene appeared as the most evolved form of Proboscidea, if not of the entire animal kingdom. As for the mammoths, which succeeded them in the Quaternary epoch and whose "rare vestiges" are found in the strata of the age of reindeer in the Vézère Valley, they were "shrunken" and already in decline and destined to disappear. The extinction of the great pachyderms must therefore be explained by the single logic of their "tendencies," since these "giants of past ages . . . were not defeated in the struggle for life." The Darwinian principle of natural selection is obviously contradicted by the disappearance of these giants, who had no predators. The transformation of the living world is no more ruled by destruction or violence than that of the earth itself. In this way, Gaudry rejected both Cuverian catastrophes and Darwinian selection in favor of a peaceful and poetic vision of nature. The "kings of nature" of each era had been herbivorous animals and not carnivores, which would have led to these destructions. Mammoths "were not the destroyers . . . much less the victims of their

The Durfort elephant in the paleontology hall at the Muséum National d'Histoire Naturelle in Paris at the start of the twentieth century. (© MNHN Paléontologie, M. Lainé photo.)

contemporaries." They became extinct only because of the necessity of their history and their tendencies. "Paleontology teaches us that we must no longer speak of fixity of species. All beings, however powerful they may be, are ephemeral. The law of change is the great law that dominates the world."[9]

Whereas Darwin thought that species are formed and vanish through competition and struggles for survival, Gaudry presented the peaceful coexistence of beings. "The geological world has not been a theater of carnage, but a theater of majesty and tranquility."[10] The terrible clashes of dinosaurs, he wrote, have been greatly exaggerated. At the end of the Miocene, "ferocious beasts" killed only to end the suffering of starving herbivores, and "except in their mating combats," animals "had no reasons to quarrel." In this way each moment of this history realized the perfection of balance and a simultaneous tendency toward progress. Gaudry picked up the Darwinian image of a genealogical tree, but less to suggest with Darwin its contingent branchings, as to conceive of its "flower" and "fruit," the perfect and ultimate goals of evolution. "This history of the living world shows us an evolution where everything is combined as if in successive transformations through which a seed becomes a magnificent tree covered with flowers and fruits, or an egg changes into a complicated and charming creature."[11] On the image of a tree is superimposed that of ontogenetic development and that of the musical notions of "harmony" and "linkages"[12] born of this dream of a continual transformation without conflicts or without really an explanation of the genealogical processes. This optimistic and progressive philosophy, which forms the basis of paleontological activity, this idyllic vision in which the conflicts were muted, seemed to animate in transformist terms the Leibnizien image of the "chain of beings" in the form of a continuous narrative.

It is a world of continuity where forms succeed and are linked together according to sometimes imperceptible transitions, in an order where "all is for the best." It is the dream of a world ruled by universal goodness. Gaudry accounted for the order and the destiny of the world through a divine "plan" that transcends it: a theodicy that, while ceaselessly marveling at the gifts a creator has given to the world, refutes the notion of evil by assigning to God the power to cause harmony to reign at each stage of this history.

When Darwin congratulated Gaudry for his "intention to examine the relations between fossil animals from the point of view of their genealogy,"[13] he probably did not want to stress the differences between his system and the French paleontologist's. But it is clear from his earliest works that Gaudry's evolutionism was deeply rooted in spiritualist philosophy.

Other French paleontologists would go even further, conceiving of evolution in spiritualist terms, often closer to those of Lamarck (orthogenesis, adaptive value of effort and habit) than those of Darwin. At the turn of the century, Darwinian theories were often criticized, and this period has often been characterized by paleontologists as a veritable "eclipse" of Darwinism.[14] In France, England, and the United States, objections to the issue of gaps in the fossil record were raised, in particular by those who rejected Darwinian materialism on philosophical grounds. How could one find in fossils the evidence for the evolutionary processes Darwin described, such as the struggle for survival and natural selection? How could the destructive power of natural selection be invoked to explain the creative processes represented by evolution and adaptation? Would it not make more sense to rather invoke some creative power of nature, some élan vital or transcendental law of change? These were the questions raised by many paleontologists between 1880 and 1920. The representations of evolutionary change as the aspect of phylogenetic trees would be profoundly modified as a result.

"Mammoth" is the right word for Henry Fairfield Osborn's monumental two-volume work on the Prosboscidea.[15] Its author was already dead by the time this pachydermic monograph was published, but his memory remains alive at the American Museum of Natural History in New York, where he founded the department of vertebrate paleontology at the turn of the century and ran it for forty years. A bronze bust of Osborn—with its plump face, impressive mustache, vest, and watch chain—stands in a corner of his restored library on the ninth floor of the museum. Next to the statue, a ticking clock slowly measures out infinite time. At the New York museum, Osborn was not only a founder but also a generous donor who brought his personal energy to paleontological research, organizing great expeditions to the American West, Africa, and Outer Mongolia in search of vertebrate fossils.

The work titled *Proboscidea* (1936–42) is actually a compilation of notes and articles by Osborn's colleagues. Osborn's work was probably to find its apogee in this monograph. Despite its uneven and unfinished quality, one can read in the book a very special attention paid to the problems of evolutionary classification. In his view of evolution, Osborn was strongly influenced by the German morphological tradition and by the thinking of the great paleontologist Edward Drinker Cope, whose disciple he was at the start of his career. In his approach to fossils, he borrowed some of the evolutionary principles Cope inherited from Lamarck, such as that of the "kinetogenesis," which supposes a link between the development or shrinking of an organ and its use.

In 1895 Osborn began to work out his own concept of evolution, apart from the "neo-Lamarckians" and "neo-Darwinists." In his eyes, evolution was founded on the interrelation of four essential factors: heredity, environment, ontogeny, and selection. Here, "selection" was but one factor among others and was mainly responsible for extinctions; it was not the motor driving evolution. This "tetra-plastic" theory of evolution was the basis for evolutionary schemas of groups of vertebrates. Osborn also brought in numerous "laws" that directed the evolution of species in an orthogenesis. Evolution proceeded by adaptive radiation, evolutionary parallelism, and morphological divergence. Osborn defined adaptive radiation as "the development of widely divergent forms in animals ancestrally of the same stock or of related stocks, as a result of bodily adaptation to widely different environments."[16] Most groups of fossil mammoths have diverged from a hypothetical common ancestor early in their evolutionary history, and thereafter followed parallel, orthogenetic lines of descent.[17]

In his day Osborn was not alone in accepting such principles. In France the paleontologist Charles Depéret also conceived of evolution as an "orthogenesis" ruled by transcendent "laws," such as the tendency to grow or shrink in size. For Depéret, characteristics shared by species in different lineages can only be independent acquisitions, or "parallelisms." Only unspecialized "primitive" forms can engender new lines. Once the "branches" have reached a certain degree of specialization in their structure, they can continue to evolve only in the same direction toward more specialized life-forms, or become extinct. Strictly applied, this principle makes evolutionary branchings almost impossible.

In a 1907 book whose title, *Les Transformations du monde animal* (Transformations of the animal world),[18] seems to parody that of Gaudry's great work, Depéret invoked this principle of "the irreversibility of evolution" to reject the attempt to bring to light any "links" between fossil genera. The supposed filiations were founded on illusory resemblances, and one had often seen "passage forms" between mistakenly grouped genera. Wrote Depéret: "Pleading the insufficiency of paleontological documents, as has been done since Darwin, is no longer enough. Forms of passage between genera not only do not exist, but cannot have existed. Observed facts prove that these were branches, each of which had an independent evolution and history."[19]

Depéret and Lucien Mayet described the phylogeny of the Proboscidea according to those evolutionary principles in a 1923 monograph devoted to Pliocene elephants.[20] They presented "five large parallel evolutionary groups, a few of which consist of two or several branches." Each group had an independent evolutionary history, so one may assume that both fossil lines evolved separately, starting from separate ancestors. The mammoths (the third group of proboscideans named) are "characterized by their strongly spiraling tusks, skulls that are flattened from front to back . . . flat foreheads, large molars with regular enamel bands." Moreover, one must also distinguish "two, and probably three, parallel branches" in this group: the *Elephas trogontherii* branch, which disappeared in the beginning of the Quaternary without leaving descendants; the *Elephas astensis primigenius* branch, which "extends unchanged from the late Pliocene to the end of the Quaternary"; and the Siberian mammoth branch (*Elephas primigenius sibiricus*), characterized by a "thick coat and long hairs, as shown by the discovery of frozen carcasses in Siberia and drawings made by men at the end of the Paleolithic." It would be wrong to think, as Gaudry and other Darwinian paleontologists did, that some of these forms were born of the others, and that the "Siberian mammoth" could have derived from the "normal mammoth" by the "gradual increase in the number of molar plates, which would then become more tightly spaced and with thinner enamel." This hypothesis, though it confirmed a "law" of the transformation of elephantine molars, was contradicted by the fact that only Siberian mammoths have thick coats. To Depéret and Mayet, it was inconceivable that a "normal mammoth" living in southern Europe could have acquired fur by migrating toward a colder climate. "That explanation is completely childish and unacceptable," they wrote. "If a hairless mammoth had been subjected to a gradual cooling which became as intense as the one shown by the Quaternary history of Europe, one of two things would have happened: it would either have died in place or migrated southward without growing hair." Specialization forbids the acquisition of new characters. According to these authors, one must therefore conclude that the Siberian mammoth had an unknown and independent origin, which should probably be sought in northern Eurasia.

It was thus no longer possible to construct a phylogenetic "tree," but only a table in which the names of species succeed each other in parallel lines whose origins are lost in the unknown. This was a paradoxical representation of evolution; by multiplying parallelisms, it made the notion of filiation inconceivable.

This was also Osborn's view, and his phylogeny of the Proboscidea is a particularly revealing application of his ideas on evolution. The first years of the twentieth century brought important discoveries about the roots of the Proboscidea family tree. In April 1901 British Museum paleontologist Charles W. Andrews and geologist Hugh Beadnell discovered two very primitive forms of Proboscidea in Africa, in the Egyptian El Faiyûm deposits: *Moeritherium* in the late Eocene period and *Palaeomastodon* in the fluvio-marine deposits of the early Oligocene.[21] According to Andrews, these were two ancestral forms of the Proboscidea order, with the older one (*Moeritherium*) giving rise to the more recent (*Palaeomastodon*), which in turn was the ancestor of the European mastodons of the early Miocene. In 1907 Osborn sent an expedition to El Faiyûm that brought back new remains of these two fossil species. Based on certain anatomical and behavioral characteristics—the first species lived on the shore, whereas the second lived in the water—he concluded that they belonged to different lineages. There was no way to consider *Moeritherium* as the ancestor of *Palaeomastodon*. They probably derived from the same unknown ancestor but had to be arranged on distinct evolutionary pathways. Like Depéret, Osborn conceived of a polyphyletism that split these two Proboscidea forms at a very early epoch.

In Osborn's 1936 monograph, the evolutionary "tree" of the proboscidean family is extraordinarily enlarged at its base and consists practically only of parallel independent branches.[22] The three superfamilies (Elephantoidea, Stegodontoidea, and Mastondontoidea) are subdivided into a great number of parallel branches that only meet deep in the Eocene and Oligocene. Even though Osborn noted an abundance of mastodons in the Tertiary, he refused to consider the hypothesis of a possible descent and considered that the Quaternary elephants were born from hypothetical ancestral forms. As with Depéret, the tree has practically no branching. Each of the groups appears separately without any known intermediary stage, from a supposed unknown ancestor in the early Eocene or the late Cretaceous. The "adaptive radiation" explodes in a multitude of lines that converge only in the very ancient common ancestor from which they all issue. The evolutionary mechanisms are linear processes oriented toward an end, orthogenesis, "rectigradations," ruled by the morphological laws of evolution. The classification of the Proboscidea also emphasized other "orthogenetic" tendencies, such as the law of a tendency to an increase in size. Elephants become bigger and bigger, then shrink and become extinct. The evolution of species follows the development of the individual: growth, aging, and death.[23]

Though dominant in the United States at the beginning of the century, Osborn's work was out-of-date by 1930. Because of his insistence on the phenomena of orthogenesis and the multiplication of classes and his refusal to accept genetics, his thinking no longer corresponded to the major scientific interests of his time. But Osborn's work as an "entrepreneur" at the American Museum of Natural History strengthened the prestige of American paleontology (centered at the New York museum) and gave research a powerful dynamism during the first half of the twentieth century. The generation that followed would shape its scientific ideas in critical reaction to Osborn's work.

During the twentieth century, a major renewal in views about the phylogenetic history of species would arise precisely from the encounter between paleontology and genetics. The discovery of the laws and mechanisms of heredity would shed new light on the question of the causes and processes of diversification. Evolution would no longer be read only in the morphological transformations of living beings. The mechanisms that control heredity in successive generations obey statistical laws and are explained at the level of the gene.

The year 1900 marked a veritable revolution in biological thinking. In that year three botanists—Hugo de Vries, Karl Erich Correns, and Erich Tschermak—separately rediscovered the work of the Czech monk Gregor Mendel on the hybridization of peas, which had led him to formulate the statistical laws of the inheritance of characteristics. What they mainly redis-

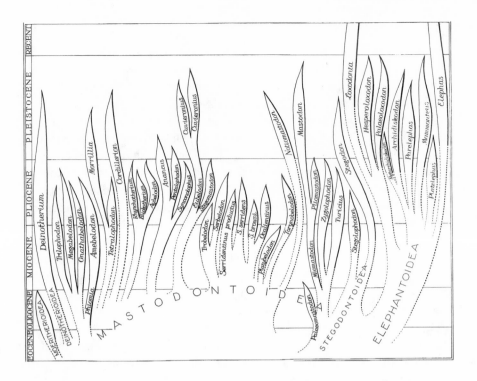

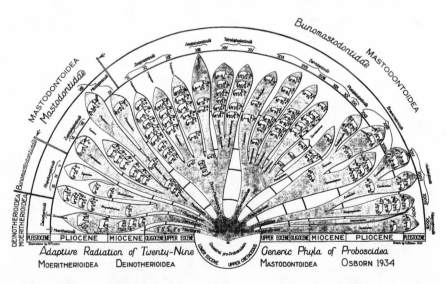

Evolutionary schema of the proboscideans and radiation of the
Mastodontoidea, according to H. F. Osborn, *Proboscidaea* (1936).

covered was the "attitude" of Mendel himself, who had brought quantification to the biological sciences. "With Mendel, biological phenomena suddenly acquired the rigor of mathematics. Statistical treatment and symbolic representation imposed a new internal logic on heredity," writes François Jacob.[24] The first geneticists, like de Vries and Gregory Bateson, were "mutationists." They believed that new species appeared after abrupt changes determined by significant "mutations," which are oriented in a given direction within the constraints inherent in the mechanism of development. In denying that evolution took place gradually through the accumulation of small variations that give rise to natural selection, the "mutationists" tended to reject Darwin's thinking, and even proclaimed the "death" of Darwinism.[25]

The following generation would engage in major debates over these theories. On the one hand, the works of geneticists led to a revision of the notion of mutation. Thomas H. Morgan's school showed that it was possible for a very large number of different mutations to accumulate within a given species in just a few years. A great number of mutations affect the gene, combine with each other, and are often barely detectable. Morgan's new concept of "mutations" is very close to the Darwinian concept of variation. After 1930 the works of Ronald A. Fisher, Sewall Wright, and J. B. S. Haldane established a mathematical theory of natural selection. Despite differences in their concepts and methods, the three geneticists recognized the power of natural selection to change the frequency of genes within populations and the capacity for mutations to produce such variations. Unlike de Vries and Bateson, these scholars believed that evolution takes place through the selective accumulation of small genetic differences.

This approach gave new importance to the role of natural selection and raised the possibility of integrating the mechanisms that rule the heredity of living creatures with the Darwinian principles of evolution. These conclusions made it possible, for the first time, to reach agreement between laboratory biologists and field naturalists, and opened a new possibility of collaboration between two groups of scholars whose approaches up to then had been very different. In this, the work of geneticist Theodosius Dobzhansky, who was trained in the first decades of the century in Sergei Chetverikov's laboratory in Russia, had a very important impact. In his 1937 book *Genetics and the Origin of Species*, Dobzhansky devoted a chapter to natural selection, presenting it not as an abstract theory, but as a process than can be proven experimentally. His work also led to the development of the new concept of population. "Evolution is a change in the genetic composition of populations," wrote Dobzhansky.[26] The typological notion of species, which had been used in experimental genetics, was criticized in favor of the notion of population. As Ernst Mayr explains, "A real understanding of natural selection, specialization and adaptation was not possible until the notion of population replaced that of typological species."[27]

The encounter between geneticists and naturalists was called the New Synthesis. This expression appears for the first time in the seminal 1942 work by Julian Huxley, *Evolution: The Modern Synthesis*. It was born in the United States in informal meetings of geneticists, embryologists, paleontologists, and systematicians between 1930 and 1940, as a common ground between different specialists in evolution. It was also known as the Neo-Darwinian Synthesis, which indicated that it retained and refined most of Darwinian evolutionary concepts, but, following August Weismann, rejected the notion of the inheritance of acquired characters that Darwin himself had accepted. The principal actors in this theoretical renewal of Darwinism were the American biologists and naturalists Ernst Mayr, Glenn Jepsen, and George Gaylord Simpson. This collaboration of specialists from previously distinct domains led to a profound intellectual and institutional reorganization of the life sciences and was expressed by many collective publications, notably the 1949 proceedings of a colloquium that had taken place two years before at Princeton University.[28] In 1947 an important conference entitled "Paleontology and Transformism" took place at the Sorbonne in Paris, furthering dialogue between European and American paleontologists and geneticists.

The New Synthesis appeared as a theory of genetic change, applied to every aspect of evolution. The gradual evolution of species could be explained by the appearance of "mutations," small accidental changes at the level of the gene, which are then sorted by natural selection. All evolutionary phenomena are the result of the same genetic mechanisms. At the geological level, fossils are proof of this transformation. The paleontologist George Gaylord Simpson was the great theoretician of the application of the principles of the New Synthesis to paleontology. His 1944 book *Tempo and Mode in Evolution* first proposed an approach to the particular processes of "macroevolution" founded on the quantitative approaches to the genetics of populations. Simpson undertook to calculate the rate of evolution. "How fast, as a matter of fact, do animals evolve in nature? This is the fundamental observational problem of tempo in evolution. It is the first question that the geneticist asks of the paleontologist."[29] The "determinates of evolution" approach (variation, mutation, length of generations, size of populations, and selection) also gives rise to quantified evaluations. As Stephen Jay Gould writes, "*Tempo and Mode* contains 36 figures, but only one portrays an animal. . . . The rest are graphs, frequency distributions and pictorial models. No paleontological innovation could have been more stunning than this."[30]

Based on these calculations, Simpson proposed three models for the evolution of a species, according to three kinds of processes: phyletic continuity, which consists of the "sustained directional (but not necessarily rectilinear) shift of the average characters of populations"; speciation, which is "the local differentiation of two or more groups within a more widespread population"; and quantum evolution, which proceeds by a "rapid shift of a biotic population in disequilibrium to an equilibrium distinctly unlike an ancestral condition."[31]

With evolution brought back to material mechanisms and quantifiable principles—and no longer to supposed "laws" of morphological change governed by such vaguely metaphysical notions as élan vital or the "tendency toward progress"—it was now possible to reconsider concepts developed by earlier paleontologists. First among these was the teleological representation of the living world, organized hierarchically toward a goal that had long made it possible to give a "direction" to evolution, in viewing the evolutionary approach to the animal kingdom, usually conceived as destined to culminate in the king of nature—man. It was also possible to account for phenomena of "orthogenesis" or straight-line evolution, without bringing in mechanisms other than those of natural selection.

Another supposed "law of evolutionary transformation" also came under fire, Louis Dollo's famous "doctrine of the irreversibility of evolution." Paleontologist Alfred Romer explained: "Such a 'doctrine' has made it impossible to constitute reasonable phylogenies for different groups for which we have a great quantity of fossil material, and it has been responsible in many cases for the assumption that successive representatives of a group did not descend from each other."[32] As we saw with Depéret and Osborn, this doctrine had in fact led to the construction of phylogenies in the form of parallel "ladders."

It was also possible to suggest new phylogenetic constructions in which the branchings represented genealogical relations between species. A famous example of this "phylogenetic tree," with its complex and unpredictable branchings, appears in the evolutionary schema of the Equidae family. Unlike the linear progression drawn by Edward Drinker Cope at the end of the nineteenth century, Simpson's phylogenetic tree presents a sheaf of lines diverging from successive "radiations." This translates the complex modalities of the evolution of species, where "the differences of rates and the quality and direction of evolutionary modifications are evident."[33]

In 1945 Simpson drew the phylogenetic tree of the Proboscidea showing that the two groups of mastodons and elephants "subtly interpenetrate each other."[34] He suggested eliminating the subdivision that Osborn had defended and bringing the two groups together in a "superfamily," the Elephantoidea. This new classification of the proboscideans was picked up in 1969 by the Austrian paleontologist Erich Thenius.[35] The genealogical tree of the Proboscidea that he built defines four large classes (Moeritherioidea, Barytherioidea, Dinotherioidea, Elephantoidea) and shows the possible roots of the Elephantidae in a Tertiary mastodon layer. Thanks to a new representation of evolutionary processes, it became possible to draw the genealogical branchings of the tree of life, whose complexity remains to be discovered by paleontological work.

Following the Second World War, the mammoth found itself in the center of a cluster of converging biological disciplines, all contributing to the reconstruction of past life. By their partial and fragmentary nature, fossil remains often raise more questions than they answer, but they alone can yield the history of living organisms through their succession, diversification, and complexification. The diversity of species through time is in itself proof of evolution, and fossils are essential to the knowledge and understanding of that history. Their study allows one to identify evolutionary lines and the processes and trends of evolution revealed by these phyletic series. "Such series are the factual material against which any theoretical conclusion must be checked," wrote Romer.[36] Fossils may no longer appear as the only "proofs" of evolution, nor do they allow us to discern its most delicate mechanisms, but they are essential to the knowledge of life itself, in all its diversity and complexity, in the detailed anatomy of organisms and their succession, and the study of their living conditions and their environment. As in Cuvier's time, knowledge of the history of life is still needed to reconstruct animals from the past, gather their remains, and ascertain their geographical distribution. The knowledge of the evolution of species appears in the details of their history and cannot be reduced to general principles.

To know the mammoth as it existed in time and space requires excavation, description, reconstitution, and classification. It demands a geological and stratigraphical approach backed by dating techniques. It requires knowledge of embryological development, which gives evolution its general orientation, and a taphonomic or ecological approach, which reconstitutes fauna and the living conditions of extinct species. Finally, only the study of fossils lets one ask the essential question of the interplay of micro- and macroevolution. The evolutionary history of organisms at the geological level cannot be reduced to a study of evolutionary mechanisms at the cellular level. It has its own tempos and modes, and it raises other questions, other parameters, and other forms of causality.

Starting in the 1970s, a new generation of American paleontologists reclaimed the specificity of questions raised by paleontology and renewed their approach to evolution at the geological level. These scientists critically reexamined Darwin's gradualism, which had inspired the New Synthesis, and

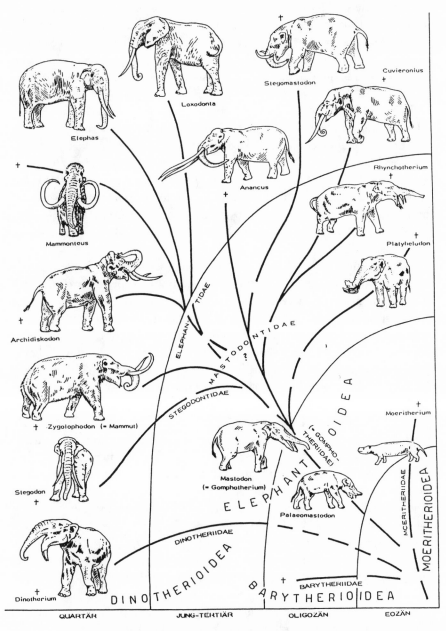

The proboscidean evolutionary tree, according to Erich Thenius, "Philogenie der Mammalia," in *Handbuch der Zoologie* (1969), p. 196. (© 1969 De Gruyter, Berlin.)

rethought the structure of "macroevolutionary" scenarios, by raising the possibility of abrupt changes, veritable "revolutions" of life. Even in Darwin, the idea that *natura non facit saltus* (nature doesn't make leaps) may have been a prejudice rooted in old philosophical beliefs and in a "progressive" view of history linked to an ideological model inspired by the social values of Victorian England.

In their 1972 manifesto of the theory of "punctuated equilibria,"[37] Niles Eldredge and Stephen Jay Gould attacked the prejudices underlying the gradualist representations of the paleontological transformation of species: "Paleontology's view of speciation has been dominated by the picture of 'phyletic gradualism,' " they wrote. "It holds that new species arise from the slow and steady transformation of entire populations. Under its influence, we seek unbroken fossil series linking two forms by insensible gradation as the only complete mirror of Darwinian processes; we ascribe all breaks to imperfections in the record."[38]

Eldredge and Gould claimed that evolution proceeds not by phyletic continuity (the slow and progressive transformation of one species into another), but to the contrary by successive speciations (formation by reproductive isolation of species that are different from the ancestral species) at particular moments that favor the explosion of evolutionary possibilities. In which case, the changes are rapid and the intermediate forms that represent this phenomenon are limited to a short period, and in small populations. So if these intermediary "links" are missing, it is because they do not exist or are very rare. "If new species appear very quickly in small, peripherally isolated local populations," Eldredge and Gould wrote, "then the great expectation of insensibly graded fossil sequences is a chimera. A new species does not evolve in the area of its ancestors; it does not arise from the slow transformation of all of its forebears. Many breaks in the fossil record are real."[39] Moreover, evolution does not necessarily move in the "optimistic" (and anthropocentric) direction of continually improved adaptation, with the best adapted species surviving while others disappear.[40] The evolution and extinction of species can be determined by accidental and contingent events.

This came as a challenge to an image inherited from the tradition of the nineteenth century and largely accepted by the supporters of the New Synthesis: that of slow change taking place through the gradual accumulation of tiny modifications. Within this new intellectual framework, the notion of a linear and progressive evolution was being challenged. The "gaps" in the geological archives that Darwin complained of could well be part of the history of life itself, representing an evolution that proceeds by sequences of long evolutionary stages and sudden evolutionary flourishings linked to abrupt and unexpected changes in the environment. This made it possible to change the

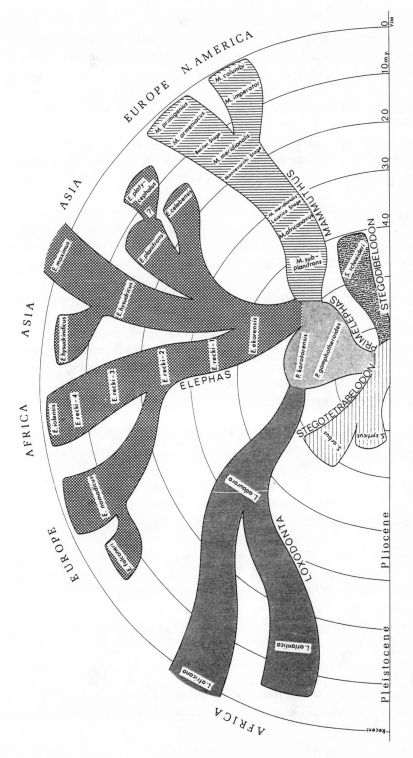

"Proposed phylogenetic relationships between all species of the family Elephantidae recognized as valid in the present study," by Maglio (1973). In *Evolution of African Mammals*, ed. Maglio and Cooke (1978), p. 361.

construction of the phylogenetic tree of the Proboscidea. A classical schema of the relationships of the genus *Mammuthus*[41] presents the succession of the three forms, *meridionalis, trogontherii,* and *primigenius,* from the early to the late Pleistocene, as a gradual transformation. *Mammuthus meridionalis* would have evolved into *trogontherii* and finally into *primigenius.* During the temporal succession of these three species, certain anatomical tendencies would have become accentuated: a change in body size, a raising and shortening of the skull, a thinning of dental enamel, and an increase in the number of enamel molar plates. When presented that way, the succession of European species appears to be a slow and progressive evolution culminating in the "hyperspecialized" form of the mammoth.

But how sure can we be that evolution occurred according to this linear schema? The scattering of remains across space and time leaves open the possibility of other hypotheses. One could, for example, suggest another reading of the remains that is not linked to gradualist assumptions. As the English paleontologist Adrian Lister notes in his revision of the classification of the genus *Mammuthus,*[42] the scattering through time of fossil evidence now available to scientists should suggest considerable caution in establishing a phylogeny. "Because of the many aspects of the morphologically intermediate character of *E. trogontherii,* it has often been considered that the change was gradual throughout the sequence. . . . And yet the three principal specimens are from .5 to 1 million years apart, and allow us to draw the schema of this change only in the broadest outline."

One must beware of the prejudices and habits of mind that lead one to interpret fossil sequences as proving an evolutionary continuity—and often lead paleontologists into erroneous gradualist interpretations. Writes Lister:

> The persistent belief in a gradual and progressive change throughout the mammoth lineage was founded on a combination of several factors: (1) The hypothesis of a gradual evolution even in the absence of evidence; . . . (2) The selection as "intermediates" of specimens that are located at the lower limit of *Primigenius* variability; (3) That "primitive" and "evolved" specimens must be respectively more recent or later; (4) The interpretation of lamellar frequency without consideration of size, so that larger specimens appear to be more primitive and smaller ones more evolved.[43]

The hypothesis of gradual change remains credible, but the possibility exists "that between two specimens there may have been a speciation event, with the last specimen having evolved as a species isolated in its reproduction, while the preceding one disappeared." In this way the image of relationships would change from one of linear continuity into a branching schema.

We have come a long way from Albert Gaudry's project of shedding light on a "divine plan" at work in the progressive history of life. The approach of the evolution of life has no other goal but that of reconstructing the events of the past in time and space, in their inevitable contingency, while taking into account both the necessity of its laws and the accidental nature of events that determine evolution. It demands that this history be told not in the shape of a general abstract or logical model, imposing an image from the outside, but in the more modest form of a "scenario," a narrative hypothesis that may or may not be confirmed by further research. The image of a bush—even more so than that of a tree, of which it is a variation—is a good model to represent the history of life in the concrete succession of its branchings, with its accidental events, its contingency, and its diversity.

From Africa to Alaska:
The Travels of
the Mammoth

"Gold! Gold! Gold! Gold!"

This headline, covering the entire front page of the July 17, 1897, *Seattle Post-Intelligencer*, announced the arrival of two ships from Alaska, the *Excelsior* and the *Portland*. The *Portland* docked in Seattle carrying sixty-eight men loaded with heavy sacks of gold from the Klondike mines. The *Excelsior* had reached San Francisco three days earlier, and fifteen of its passengers, bearded pioneers in miners' clothes, were escorted by the crowd to a bank where their gold was weighed. "Thousands of people who greeted them on the docks were suddenly seized with Klondike fever, and this evening the name 'Klondike' is on the lips of every man, woman and child in the city," wrote a reporter for the *Seattle Post-Intelligencer*.[1]

Jack London's *Call of the Wild*[2] and Charlie Chaplin's *Gold Rush* told the heroic legends and heart-wrenching tales of this second gold rush to the far north. Half a century after that of the far west, it sent thousands of adventurers and fortune seekers to the inhospitable land, frozen roads, and muddy rivers of Alaska, all bound for the new El Dorados in the Yukon Valley, the Klondike, and the Copper River. The result of this frenetic quest, on which more people set out than for the Crusades, was a profound transformation of the north and its landscape. It accelerated the exploration of the interior of a region that until then had been almost exclusively inhabited by native Indians and Eskimos. Within months, tent and then log cabin cities sprang up like mushrooms, with names that still have a mythic ring today: Dawson City, Nome, Fairbanks. It spurred the development of a fleet of steamers on the Yukon and adjacent rivers, baptized with such sweet names as *Alice*, *Sara*, and *Susie*. It required the construction of an entire infrastructure of roads and railroad lines under very difficult conditions. The famed White Pass and Yukon Railroad—the train that links Skagway to Whitehorse in Canada—

is the expression of a determined will to penetrate the interior, where the precious gold fields lay. Finally, the rush for Klondike gold marked the beginning of large-scale geological exploration and research about the mammoth in the frozen regions of the American north.

In western Europe and Russia, mining has played a major role in the discovery and study of fossils for centuries. In America in the middle of the nineteenth century, the gold rush was followed by a veritable "bone rush" when immense deposits of dinosaur fossils were discovered in the West. One cannot overemphasize the benefits American paleontology drew from this thirst for gold, which drove thousands to undertake a journey beyond the Rocky Mountains to dig the soil and pan the river sediments. Some of the fossil hunters that Cope and Marsh hired had started out as gold prospectors, and the famous transcontinental railroad, built to link one end of the country with the other, played an essential role in transporting the enormous and precious remains of triceratops, tyrannosaurs, and brontosaurs discovered in Utah or Nevada to the more civilized climes of New England or New Jersey, where dignified scientists indulged their passion for collecting and studying.

The Alaska gold rush was also the occasion of important paleontological explorations. At the beginning of the century, intensive mining had a decisive influence on the study of the territory's geology and paleontology. The main goal was to evaluate the mineral wealth of an area that was still poorly known—though thought to be "of great economic importance"[3]—but also to better understand the prehistory of the region, which was rich in fossil remains.

Expeditions led by serious scientists followed in the footsteps of the sourdoughs and the "chichacos" (an Indian word meaning "tenderfoot") who dug the riverbed gold veins. At the turn of the century, several expeditions were sent from Canada and the United States with the twin goals of scientifically mapping the gold fields and other mineral resources and of exploring fossil sites—perhaps to bring back mammoth skeletons or their mummified carcasses.

The history of the first discoveries by westerners of mammoth remains in Alaska overlaps that of the search for the legendary Northwest Passage. At the beginning of the nineteenth century, European explorers seeking a route to Europe through the chaos of icebergs and the labyrinth of frozen seas of the great Arctic north saw mammoth bones and tusks on the west coast of Alaska. Otto von Kotzebue gave the first account of this memorable discovery in his *Voyage of Discovery into the South Sea and Beering's Straits in the Years 1815–1818*.

August the 8th. We had passed a very unpleasant night, for it was stormy and rainy; and as the morning promised no better weather, I resolved to sail back to the ship; but scarcely had we gone half way, when we were overtaken by a violent storm from the south-east; the long-boat drew much water, and we were obliged to return to the landing-place we had just quitted. . . . It seemed as if fortune had sent this storm, to enable us to make a very remarkable discovery, which we owe to Dr. Esch-scholtz. We had climbed much about during our stay, without discovering that we were on real ice-bergs. The doctor, who had extended his excursions, found part of the bank broken down, and saw, to his astonishment, that the interior of the mountain consisted of pure ice. At this news, we all went, provided with shovels and crows, to examine this phenomenon more closely. . . . We saw masses of the purest ice, of the height of an hundred feet, which are under a cover of moss and grass; and could not have been produced, but by some terrible revolution. The place which, by some accident, had fallen in, and is now exposed to the sun and air, melts away, and a good deal of water flows into the sea. An indisputable proof that what we saw was real ice is the quantity of mammoths' teeth and bones, which were exposed to view by the melting, and among which I myself found a very fine tooth.[4]

In honor of his discovery, the name Eschscholtz was given to the bay and that of Kotzebue to a nearby sound. In 1816 an English explorer named Captain Frederick Beechey also dug up fossil elephant remains on the Alaskan coast in Eschscholtz Bay during a voyage to the Bering Strait. These bones, which were gathered by the expedition's surgeon, were sent to the geologist and minister William Buckland for examination. His analysis was published in 1831 in an appendix to the account of Captain Beechey's voyage. Buckland identified the remains as those of the animals that Cuvier had named "fossil elephants" and—as a good orthodox Cuvierian—maintained that whatever had killed the mammoths could only have been catastrophic: the mammoths had suddenly died of cold. Their successful adaptation to polar climates and the thickness of their woolly coats had been exaggerated, he said. Buckland also rejected the idea that the fossils could have been trapped in icebergs.

The cliffs containing the bones, which have been described by Kotzebue and Eschscholtz as icebergs covered with moss and grass, are not composed of pure ice, but are merely one of the

Collecting mammoth tusks in Kotzebue Sound, Alaska. Lithograph by Thomas Woodward, ca. 1850. (Anchorage Museum of History and Art photo.)

ordinary deposits of mud and gravel, that occur on many parts of the shores of the Polar Sea, being identical in age and character with Diluvial deposits of the same kind which are known to be dispersed over the whole of Europe, and over a large part of Northern Asia and North America.[5]

Expeditions to western Alaska took place throughout the nineteenth century. In 1881 the *Report of the Crossing of the U.S. Revenue Steamer Corwin in the Arctic Ocean in 1880* discussed the "glacier" theory that Kotzebue had defended and described the discovery of a great number of mammoth bones and tusks and of a few smaller bones, probably those of aurochs and musk oxen, at a place called Elephant Point "in Eschscholtz Bay, about fifteen miles from Buckland River." In the same year the naturalist W. H. Dall explored sites on Alaska's west coast and studied its fossil-bearing deposits. He reported that they gave off a powerful ammoniacal smell of decomposing animal matter and said that the fact that these organic remains were preserved in ice surely played an essential role in their conservation.[6]

Around the turn of the century, new expeditions brought fresh elements to the understanding of the geological history of Alaska. Beginning in 1899 the geologist A. G. Maddren explored the Yukon Valley and various sites of the Bering Sea, eastern Siberia, and the Arctic Ocean as far as Cape Beaufort. In 1904 the Department of Geology of the Smithsonian Institution put Maddren in charge of an expedition to find complete skeletons of mammoths and extinct animals whose numerous remains had already been gathered in those regions.[7] He found several fine specimens, in particular a complete mammoth skull with a pair of impressive tusks. But his research mainly centered on a geological study of Pleistocene deposits of northern Alaska, where Kotzebue's expedition had first discovered fossil mammoth remains almost a hundred years before, and focused on the problem of distinguishing fossil-bearing Pleistocene sediments from more recent glacial deposits.

A second expedition, led by Charles W. Gilmore, set out in the spring of 1907. Gilmore was given the following instructions: "You are hereby authorized to proceed to Alaska, on or about May 22, 1907, for the purpose of exploring the regions herein described, with a view to securing remains of large extinct vertebrate animals and investigating the causes which have led to their extinction."[8]

Starting in Washington State, the naturalists stopped over in Seattle, headed for Whitehorse by way of Skagway, then descended the Yukon River to an area called the "bone yard," where many fossils had been found, and on to the Canadian interior. Along the way they spoke to miners, explored sites, and asked to see prehistoric remains found in such famous gold-mining locales as Bonanza Creek, from which many known specimens had been obtained.

> Scattered remains of Pleistocene mammals are commonly found in the diggings of this region, but the results of diligent inquiry regarding the finding of complete or partial skeletons in the mining operations conducted here were not encouraging. In only one instance were we told of the finding of an accumulation of bones such as would lead one to believe that an entire skeleton or any considerable part of a skeleton of a single individual had been found. The single case mentioned was that of the remains of a mammoth (*Elephas primigenius*) disinterred while sinking a shaft on Quartz Creek in March, 1904. The skull and tusks were recovered intact but although surrounded by a mass of other bones no attempt had been made to preserve them.

The members of the expedition found pieces of tusks, teeth, and skeleton fragments uncovered by water erosion or mining activity. "The scattered re-

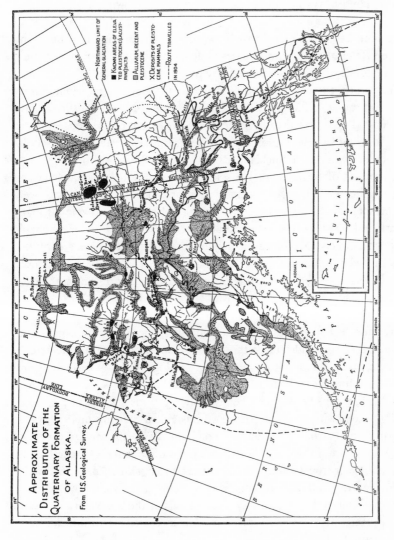

Map of Alaska showing the Quaternary geological formations and the distribution of fossil Pleistocene mammals. This U.S. Geological Survey map also shows (*dotted line*) the route of Maddren's 1904 geological expedition.

mains of Pleistocene animals occur throughout the unglaciated region of Alaska and adjacent Canadian territories." They accumulated in the black muck in gulches and stream valleys, in the fine elevated clays of the Yukon Valley, and in more recent fluvial and alluvial deposits. But there were very few accumulations of mammoth bones that allowed one to affirm they had been buried in situ. Wrote Gilmore: "The writer was shown bones protruding from the face of the undisturbed beds in the Klondike region and in other instances collected specimens actually imbedded in the elevated silts along the Yukon River; they were in all cases disarticulated and scattered, and there was no evidence of an association of any of the parts found."[9]

On its return, the expedition took the Yukon River and its tributaries by steamer and canoe to Rampart, Fort Gibbon, and to Nome, on the far west Alaska coast. But it was already September and late in the season, so it was impossible to pursue the exploration of the Pacific coast sites—Eschscholtz Bay, Kotzebue Sound, or Buckland River—before the first snows.

The account of this remarkable scientific expedition is picturesque and precise and illustrated by magnificent photographs, but the expedition's overall achievement was somewhat disappointing. The scientists explored many previously unknown locations and found and collected fossils but were unable to establish a precise geological history of Alaska or bring back a complete mammoth skeleton, much less a frozen mammoth. It would be several more decades before a synthesis of the knowledge of the region's geology would be made.[10]

With the quasi-industrial exploitation of the gold sites, the second third of the twentieth century was an extraordinarily favorable time for gathering of vertebrate fossils, and tens of thousands of specimens were collected. In 1938 alone, the paleontologist Otto Geist catalogued 8,008 specimens and sent his eight tons of treasure to the American Museum of Natural History in New York, where he worked. Today his collections are still in the museum's eight-story wing built by patron Childs Frick. Starting in 1939, Geist investigated the Fairbanks region and northern Alaska for the University of Fairbanks museum. In 1948, while exploring the frozen mud of the late Pleistocene (known in the United States as the Wisconsinian) near Fairbanks, he discovered the mummified remains of a young mammoth. This mummy fragment is displayed in the gallery of mammals at the New York museum: a brown, dried, almost hairless piece of skin in the shape of a head, neck, trunk, and forefoot.

Today the search for gold lingers as a memory of Alaska's fabulous past. Paleontologists and geologists have different concerns from those of seekers after precious minerals, and the black gold carried by the Trans-Alaska pipeline from Prudhoe Bay to Valdez is a source of far greater riches than the occasional nuggets still to be dug up. Left behind are hillsides leveled by the gold miners' hydraulic jets and the rusting skeletons of enormous dredges, articulated monsters that once scraped riverbed gravels to extract the precious metal. They are now motionless hulks, whose monstrous image is reflected in the clear and peaceful waters of lakes ruffled only by the busy activity of beavers. Their deafening metallic roar has fallen silent—the last one stopped working in the 1960s—and they are now mere tourist attractions. The few remaining gold seekers who still melt the permafrost to reach gold-bearing quartz layers every summer often earn more from the ivory mammoth tusks that protrude from the Quaternary strata than from the few nuggets they are still able to find.

From time immemorial, the region's inhabitants have used fossil ivory for many purposes. Traditionally, Eskimos and Indians fashioned various utensils from mammoth ivory, such as a "dipper as large as a child's head carved out of a single piece of ivory tusk," which Dall described in 1880. On the Yukon, Indians sometimes used chunks of tusk as weights to sink their salmon nets. Pieces of mammoth ivory were shaped into sled runners, drilled with evenly

Gold miners at Willow Creek, Alaska. The jet of water melts the permafrost, occasionally uncovering the remains of Pleistocene mammals. (Anchorage Museum of History and Art photo.)

Curios made of mammoth fossil ivory from the Yukon. (C. Cohen photo.)

spaced holes, and lashed to wooden frames.[11] They were also made into jewelry—necklaces, bracelets, and earrings—and traditional "scrimshaw," ivory inscribed and colored with black or red ink.

With the mass influx of whites at the beginning of the twentieth century, ivory handicraft workers turned to producing curios such as paperweights or carved figurines. The mammoth, with its curving tusks and its "skirt" of long hairs, was a favorite subject. "One finds objects of this sort in the curio shops of Nome," wrote Gilmore at the beginning of the century. "The Skagway merchants get most of their tusks from the Klondike region, whereas those in Nome get ivory from Eschscholtz Bay, Buckland River, and the area around the Kobuk River." The people of the far north continue to express themselves artistically by carving or inscribing walrus ivory and mammoth tusk ivory, which is usually yellowish but is sometimes black or streaked with dark colors. Today mammoth bone and ivory carvers can be found from Nome to Fairbanks and Anchorage, as what was a traditional native handicraft has become part of a profitable tourist trade also plied by the white settlers.

But ivory from Alaskan mammoths, which is often rotten, discolored, and exfoliated, compares poorly to that from Siberia. From late spring until fall, Siberian natives also gather the precious tusks and make them into objects of all sorts. In Russian art collections and monasteries, one can find ancient figurines and art objects sculpted in fossil ivory. As Eugen Pfizenmayer wrote at the beginning of the twentieth century: "At one craftsman's in Yakutsk I found a model in ivory of the old Cossack fortress, and models of a Yakut summer yurt, and of sledges with an ox and reindeer harnessed to them. There were rings and combs, too, and boxes in fine fretwork. Powder horns, matchboxes, knife-handles, parts of the harness for horses and reindeer and dogs, and many other articles were also made from the tusks."[12]

The mammoth plays an important role in the art and handicraft of these peoples, as its remains are found in abundance on both sides of the Bering Strait. "American ivory has been a trade object among the Chuchkis on both shores of the Bering Strait for at least a century and perhaps a longer, unknown period. One hardly dares calculate the immense numbers of mammoths that had been buried in gravesites on the American side of the Bering Strait," wrote an English traveler named Richardson in 1854.[13]

From time out of mind, mammoth ivory has been traded between the native peoples of the American north and of the northern parts of Asia. Trade in Siberian fossil ivory with China, Persia, and Turkey has been known since the ninth century. Because of the modern ban on the sale of elephant ivory, the precious mammoth tusks from Siberia are now the object of an active and profitable commerce with Japan and the countries of Europe.

As in Alaska, scientific expeditions to Siberia in search of mammoth remains became more numerous during the nineteenth century. Many naturalists explored the geology and paleontology of northern Siberia, including Brandt, Middendorff, and de Toll. But the most famous expedition is the one that left St. Petersburg in 1901 to find the remains of a frozen mammoth on the banks of the Berezovka River. The governor of Yakutsk had alerted the St. Petersburg Academy of Sciences of the discovery of the remains of a mammoth that, according to the head of the Kolymsk district who had gone to inspect them, was very well preserved. It had been discovered about thirty-five yards above the collapsed bank of the Berezovka, a tributary of the Kolyma River, which flows to the Arctic Ocean. A Lamut hunter named Semyon Tarabykin had been hunting in the area in August 1900. He followed his dog, who, "attracted by the enticing food it smelt, led him to the mammoth body."[14] When it was found, the mammoth's head and flesh were still intact, with a trunk and a tusk visible. The head and back must have been exposed for some time, because wolves had torn off pieces of skin and flesh from several parts of the back and trunk. At the end of August, the hunters chopped off the tusks with axes and exchanged them for goods at Kolymsk from a cossack named Yavlovski. He notified the local authorities, since he knew the

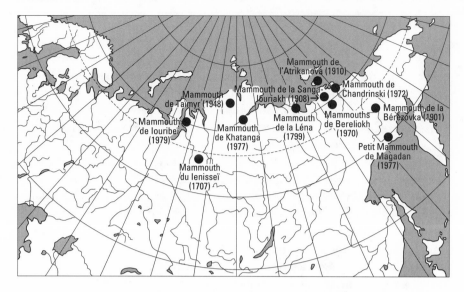

Map of Siberia showing the most important sites where mammoths have been found since the eighteenth century. (N. K. Vereshchagin.)

St. Petersburg Academy had promised rewards for the discovery of intact bodies of these prehistoric animals. The Academy decided to recover the specimen and received a government grant of 16,000 rubles for the expedition. A team of naturalists led by entomologist Otto F. Herz traveled to the site. Eugen W. Pfizenmayer, who had the crucial role of taxidermist, gave us an account of this now-legendary expedition.[15]

The expedition left St. Petersburg on May 3, 1901, for the long journey through northern Europe and Asia. At Irkutsk, the capital of central Siberia, the travelers had to abandon their luxury express train, which had, "as well as the dining-car, a saloon with a piano, for music, smoking and reading; a bath; and even a car fitted up as a church for the orthodox."[16] They continued their journey aboard a small commercial steamer, then by horseback, and finally by reindeer sled. At the end of this long voyage, which was full of interesting encounters—with native villages, a community of ascetics, gold miners, and the political exiles of Verkhoyansk—they reached the discovery site. "Some time before the mammoth body came in view I smelt its anything but pleasant odor—like the smell of a badly kept stable heavily blended with that of offal. Then, round a bend in the path, the towering skull appeared, and we stood at the grave of the diluvial monster. The body and limbs still stuck partially in the masses of earth along with which the corpse had been precipitated in a big fall from the bank of ice."[17]

The position of the mammoth revealed what had caused its death: a fall into a crevasse, from which the animal, because of the weight of its body, had not been able to free itself. "The right forefoot was doubled up and the left stretched forward as if it had struggled to rise. But its strength had apparently not been up to it. . . . [I]n its fall it had not only broken several bones, but had been almost completely buried by the falls of earth which tumbled in on it, so that it had suffocated."[18]

To dig out the animal, it was first necessary to unfreeze it. Herz's team built a log cabin over the mammoth's body, with two Yakut stoves and chimneys. The mammoth's head was much too high for the cabin roof, so it was dissected by cutting the sheaves of muscle that still connected the skull to the jaw. This revealed "half-chewed food still in its mouth, between the back teeth and on its tongue, which was in good preservation." The gradual thawing of the mammoth allowed it to be dissected with its parts removed one after another: first the skin, then sections of flesh, then the internal organs. "Brownish-black" blood clots were visible near the stomach, whose contents spilled through huge openings. Some thirty-five pounds of plant material were gathered from the mouth and stomach, making it possible to study mammoth feeding habits and Siberian vegetation of the period. The liver, heart,

The Berezovka mammoth, shortly before its excavation, fall 1901.
(V. E. Garutt personal collection.)

and lungs had disappeared, no doubt eaten by scavengers. The chunks of flesh that were successively removed from the foreleg, thigh, and pelvis were lined with thick layers of fat.

"As long as it was frozen, [the meat] had a quite fresh and healthy appearance and a dark-red color like that of frozen reindeer or horseflesh, but it was considerably coarser in fibre."[19] Once thawed, however, it looked mushy and gray and gave off a repulsive ammoniacal stench. "My readers will understand that it did not occur to us to eat a leg or shoulder of our pachyderm. We often wondered what roast mammoth would taste like, but none of us wanted to give it a try," wrote Pfizenmayer. This account should put an end to the persistent legend that this dignified team of paleontologists added mammoth meat to its menu!

Excavating the animal was not without its surprises and revealed anatomical structures that had been unknown up to then. For example, it was noted that the mammoth had four toes on each leg and not five, like those of other known proboscideans, and that it had an anal flap, no doubt an adaptation to

The members of the St. Petersburg Academy of Sciences expedition at the Berezovka site. The mammoth carcass is being thawed in a log cabin built for the purpose. (V. E. Garutt personal collection.)

cold. Under its stomach, the scientists also discovered the animal's perfectly preserved erect penis.

> Three days earlier, we were able to uncover a soft, but solidly frozen organ that we could not identify, under the foot of the left posterior leg. It stuck out somewhat below the skin of the stomach, which was also frozen. This unknown object now revealed itself as the penis, which had been completely flattened by the weight of the body resting on it. It was completely extended, measuring 3 feet long, 7.5 inches thick, and 4 inches above the orifice of the urethra. If the erection was caused as a result of the animal's difficulty in breathing, this is one more proof that it was smothered after its fall into the icy crevasse.[20]

The dissection of the mammoth indeed allowed the naturalists to confirm the causes of its death: the humerus was fractured in the middle, and there was a hematoma between the muscles, conjunctive tissue, and fat. The fracture no doubt occurred when the animal fell, because there was also a compound fracture of the pelvis with a similar pattern of blood flow.

The mammoth was cut up into pieces in a fashion adapted to both the requirements of scientific study and the means of transport: sleds pulled by pairs of reindeer. "Not to damage our prospects of making anatomical investigations later, I cut through both forelegs at the joint," wrote Pfizenmayer.[21] The pieces were sewn into cow- and horsehides with the hair facing in, then exposed to air and allowed to freeze again. "In this the severe frost, usually our greatest enemy, was our best friend. And how our anxious colleagues in St. Petersburg had racked their brains before we left to think of all the possible methods of preservation!"[22] The mammoth, almost entirely cut up and suitably packaged, was faithfully guarded at night by Yavlovski's dog.

The excavation of the mammoth took place between September 21 and October 10, 1901, making it possible to haul the precious carcass, still frozen, from the depths of Siberia in sleds pulled by dogs, then reindeer and horses, to a refrigerator car waiting near Irkutsk. It reached St. Petersburg before the end of winter, on February 18, 1902. Two days later, the hastily reassembled mammoth graced the grand entry of the St. Petersburg Zoological Institute, where it was shown to "the imperial couple, accompanied by a numerous retinue." Czar Nicholas II took a certain interest in the museum director's explanations, but the czarina, holding a handkerchief over her nose against the smell, asked if there was "something else of interest to see in the museum, preferably far away."

The stuffed and mounted Berezovka mammoth in its glass case at the St. Petersburg Zoological Institute, in the very pose in which it was found. (V. E. Garutt personal collection.)

The results of the expedition were published in an important two-volume monograph.[23] Not only were the usual paleontological skeleton and teeth available for laboratory study, but also—and this was truly fantastic for an animal more than thirty-five thousand years old—the structure of the skin, the hairy coat, a serological study of the blood, and the animal's feeding habits, and therefore its contemporary vegetation.

Today the Berezovka mammoth can still be seen at the St. Petersburg Zoological Institute, sitting in a glass case, its skin and hair restored, in the very position it was found by the Lamut hunters in 1900.

Other discoveries of frozen mammoths followed. In 1908 a new expedition was organized to retrieve a mammoth found on the banks of another Siberian river, the Sanga Yurakh. The soft parts of the body were in very poor condition, but a few fragments of the precious remains could be conserved, including the trunk, about which little was then known. In 1910 the naturalist Volosovich brought the remains of a mammoth from the Lyakhov Islands; its skeleton, a mummified foot, and a few pieces of skin are now at the Paris Muséum National d'Histoire Naturelle. On June 23, 1977, a nine-month-old baby mammoth, entirely preserved in frozen sediments, was discovered on the Magadan Peninsula in the easternmost part of Siberia. It was brought back to St. Petersburg by a team led by Nikolai Vereshchagin.[24] The specimen has been dated to 43,800 years ago, plus or minus 4,200 years. This sensational discovery made it possible to study the growth and biology of mammoths, their ecology and taphonomy, and to carry out molecular biology experiments on their flesh. Today excavation and conservation conditions have improved; the ancient remains of these animals are no longer carried by reindeer or on mule back, but in planes or by helicopter. Still, the discovery of a frozen mammoth is far from commonplace. Each time it is an exceptional event, and one likely to bring us new knowledge about life and nature in prehistoric times.

While some scientists searched for mammoths in Siberia and Alaska in the early decades of the century, others were seeking their origins in Africa. The remains scattered throughout Europe, in Italy, France, Germany, and England, had raised the question of the birthplace of the mammoth and the Proboscidea. Many looked for it in Asia, partly for anatomical reasons—Cuvier had identified the mammoth as anatomically close to the Indian elephant—partly for more or less mythological ones: Asia was thought to be

The mammoth takes the plane. The expedition's return from Taymyr, 1949. (V. E. Garutt personal collection.)

the location of the Garden of Eden and the cradle of animal and human life. But the birthplace of the mammoth—and of man—would be found in Africa.

In April 1901 the English paleontologist Charles W. Andrews of the British Museum and the geologist Hugh Beadnell found remains in the Egyptian El Faiyûm that they identified as very primitive Proboscidea. One, which according to them was amphibious, was the *Moeritherium*; the other, which they baptized *Palaeomastodon beadnelli*, was found in early Oligocene fluvio-marine deposits. The cradle of the Proboscidea seemed close at hand. In 1907 a new expedition to Africa, led by Walter Granger and George Olsen, confirmed the find and revealed other ancestral African proboscidean species.

Despite the debates that still rage over the monophyletic origin of Tertiary proboscideans, it is now clear that they were indeed born in Africa early in the Tertiary era, some 60 million years ago. They began as animals of the size of a small horse and looked somewhat like a tapir. They diversified in Africa in the Miocene some 20 million years ago, then migrated to Europe, Asia, and Africa. The unique genus *Elephas* split into three distinct genera, each with a distinct history: *Loxodon*, *Elephas*, and *Mammuthus*.[25]

The genus *Mammuthus* was actually known only in southern Europe, but it appears more and more likely that its roots were located in the African continent. In 1978 a Franco-American paleontology team attempted to trace the ancestors of the *Mammuthus* genus from newly discovered African forms, in a volume devoted to the paleontological study of African mammals. "Although this genus is generally thought of in terms of the mammoths of northern continents, like all other elephant groups this one also had its origin in Africa. The early species of the genus occurring in Africa are the succession of *M. subplanifrons*, *M. africanavus* and *M. meridionalis*."[26] The earliest form (*Mammuthus subplanifrons*) is known only from a few dozen fossil remains, mainly teeth, discovered in a few early and middle Pliocene sites (dated between 5 million and 4.5 million years ago) of South and East Africa. *Mammuthus africanavus* is the first well-defined species of the genus; its remains in Algeria and East Africa date from the middle and late Pliocene (approximately 3.2 million years ago). The third species, *Mammuthus meridionalis*, was known from a much older epoch in Europe—in the layers of the early Pliocene in Spain, Sicily, and southern France, between 3 million and 2.5 million years ago. How can one explain the presence of this form in the early Pleistocene in North Africa? Did the European *meridionalis* derive from an *africanavus* ancestor that had migrated from the other side of the Mediterranean during the Pleistocene—whereas the North African *meridionalis* represented a stage of the *africanavus* form that had evolved in place in Africa, parallel to the

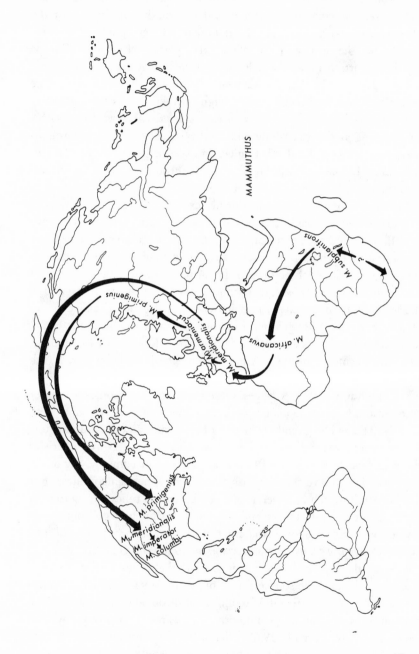

The expansion of the genus *Mammuthus* from the early Pliocene to the end of the Pleistocene, from Africa to North America. (Maglio, *Transactions of the American Philosophical Society*, n.s., 63 [1973]: 116.)

European species? Or was the African *meridionalis* the result of a reverse migration from Europe after *africanavus* became extinct in Africa? Whatever the case, *Mammuthus meridionalis* did not survive in Africa past the mid-Pleistocene and gave way to the dominant species *Loxodonta atlantica*. From then on, the history of this genus takes place exclusively in Europe and in Asia. Starting in southern Europe (*Mammuthus meridionalis*) at the end of the Pliocene, 2.5 million years ago, it extended into central Europe (*Mammuthus trogontherii*), and then through northern Eurasia (*Mammuthus primigenius*) and North Africa (*Mammuthus imperator, Mammuthus columbi*). Mammoths migrated toward Italy and southern Spain, spreading throughout western Europe to England, France, and Germany, and to central Europe, northern Russia, and Asia, and from there throughout the North American continent.

Unlike the dinosaurs, whose worldwide expansion is explained by continental drift (they occupied Pangaea before it split into Laurasia and Gondwanaland), mammoths appeared and diversified in the Tertiary and Quaternary, at a time when the continents had already separated. They were present in all of the countries of the Northern Hemisphere because they were great travelers. In their migrations, mammoths covered vast stretches of the northern plains of Europe, perhaps in search of vegetation and climates adapted to their way of life, perhaps to escape that most terrible of predators, man. An immense journey, which led them from East Africa to southern Europe, from central Europe to China, from Siberia to Japan and North America.

When and how did these voyages, these immense migrations, take place? Each scenario involves paleobiographical hypotheses that are related to both the history of living beings and the history of the earth. How did the ancestors of the mammoth reach Europe? Did they migrate from North Africa to southern Europe by the Strait of Gibraltar, or did they come by way of the Middle East? It is known that they passed onto the American continent over the Bering Strait, which has been a land bridge several times during the last twenty thousand years. But when and why did the mammoths go from northern Asia to North America? In America the mammoth produced two branches, *Mammuthus imperator* and *Mammuthus columbi*. The first was limited to the northern part of North America, and the second migrated southward as far as the extreme southern part of North America, around Central America. At Mexico's National Museum of Anthropology, one can see the almost complete skeleton of a mammoth (*Mammuthus imperator* Leidy) displayed as it was discovered in 1952 at Santa Isabel Itzapan, not far from the Aztec city of Teotihuacán. These fossil remains, dated to 11,000 years B.P., are displayed next to man-made instruments that were found nearby; the animal was no doubt killed by a hunting party and butchered in place. The same room has a fresco showing North American fauna around 10,000 B.C., in which one can recognize the heavy, dark-brown silhouette with spiraled tusks, dominating all the others.

The journeys of the mammoth strangely resemble those of the first men. Like man, the mammoth was a great wanderer. Like man, it came from an African cradle and ranged over all of Europe and northern Asia before spreading to the American continent. Perhaps the history of the mammoth is linked to that of Paleolithic man, if it is true that man was its predator. The first humans who came from Asia reached America by the Bering land bridge some eleven thousand years ago. Were they led that way while pursuing herds of mammoths? The question of the extinction of the mammoth would therefore be directly related to that of the origin of the first Americans.[27] Certain "scenarios" have told the story of a brutal "extermination," a "blitzkrieg" against these great mammals by *Homo sapiens* who had come from Asia as skillful, well-armed hunters. In fact, one finds remains of the Clovis culture in a number of hunting locations from Alaska to Mexico, its famous grooved points bearing witness to the life of these first humans inhabiting the American continent. Archaeologists have recently found flecks of dried mammoth blood

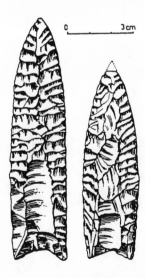

0 _____ 3 cm

Clovis points. These grooved points, made by the first inhabitants of America, were probably used to hunt and butcher mammoths. (M. Brézillon, 1977.)

on arrowheads discovered in the most ancient sites of Alaska and the western coast of North America.[28]

Today the scenarios that account for the travels of the mammoth no longer refer to Hannibal's or Alexander's expeditions, nor to the waters of the Flood sweeping animals from Africa to Siberia, as Gmelin and Pallas claimed, nor even (as Buffon suggested) a progressive cooling of the earth that caused the animals to migrate from polar regions toward more clement tropical ones.

In the history of species, the mammoths' biological evolution is closely connected to their migrations and travels. Paleobiogeography formulates hypotheses (constructs scenarios) about migrations starting from the geographic dispersal of fossil forms. It links the history of life to the history of the earth. If the wanderings of the mammoth family led it from southern Africa to the far reaches of eastern Siberia and Alaska, then reconstructing this itinerary demands that we take into account the history of the earth and its climates, its passages and straits, its islands and continental plates, its associated fauna, and the successive adaptations of species. This biogeographical study is essential to understanding the mechanisms of migrations, adaptations, speciation, and evolution. It is why we need, today more than ever, excavation expeditions and the careful search for fossil remains that mark the immense itinerary of these great vanished pachyderms.

Cloning the Mammoth?
Elephants, Computers,
and Molecules

If you step out of the Hermitage Museum and cross the Neva River on the Palace Bridge, you get a view of a row of pastel-colored palaces on the banks of Saint Basil Island. One of them is the St. Petersburg Academy of Sciences Zoological Institute, and it houses the most famous collection of mammoths in the world. In the elegant, domed hall, you can see the Berezovka River mammoth in its glass case, sitting as it was found in 1901, trapped in a crevasse. Marching down the center of the hall is an impressive procession: the skeleton of the Lena River mammoth found in 1799—the first specimen discovered by Western scientists, frozen and still covered with hair; the skeleton of another mammoth dug up near the Berezovka; and that of *Archidiskodon meridionalis*, the biggest mammoth in the world preserved in a museum. But the high point of this collection is undoubtedly the mammoth discovered on the Taymyr Peninsula in eastern Siberia in the fall of 1948. The skeleton of this precious, somewhat small specimen (8.5 feet at the shoulder) is almost complete. It was remarkably well reconstructed in 1950 under the direction of St. Petersburg Zoological Institute professor Vadim Evgenievich Garutt, a specialist in the anatomy and the classification of fossil proboscideans.[1] In 1990 it officially became the "neotype" of the genus to which the mammoth belongs, the "reference mammoth," to which all anatomical study and classification must refer.[2] In this role, it supplanted the skeleton dug up near Burgtonna in 1695, on which Blumenbach had founded the first definition of the fossil species *Elephas primigenius* at the end of the eighteenth century.

It was Blumenbach who in 1799 first gave the mammoth a Latin name in accordance with the rules of Linnaean nomenclature.[3] With this name, *Elephas primigenius* (the first-born elephant), he included the species *primigenius* in the genus *Elephas* and characterized it as "the remains of an enormously large animal," whose bones were found in large quantities in the soil of

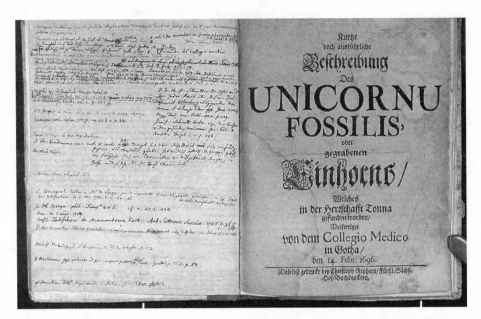

The "birth" of the mammoth: Blumenbach's notes on fossil elephants handwritten on a collection of pamphlets dealing with the interpretation of these remains. (With permission of S. J. Gould. Photo C. Cohen.)

Germany. As a type specimen, he took the Tonna elephant[4]—after having carefully studied the different interpretations of its remains—as a "sport of nature," a unicorn, or a fossil elephant. Blumenbach's notes, handwritten on the flyleaf of a collection of seventeenth-century pamphlets,[5] reveal the interest he had in the fossil. This precious collection, gathered and annotated by Blumenbach, today belongs to Harvard paleontologist Stephen Jay Gould, a great theoretician of evolution and an enlightened bibliophile.[6]

The Tonna specimen, which has since been restudied in light of better knowledge of fossil elephants, is no longer considered to belong to the mammoth lineage (*Mammuthus*), but rather to that of the Indian elephant (*Elephas antiquus*). This is why a neotype has been officially designated. According to the strict rules of the International Code of Zoological Nomenclature,[7] an animal must retain the original species designation given to it by its first investigator, whose name is associated with it, along with the date of its first publication. For this reason the mammoth was long known as *Elephas primigenius* Blumenbach 1799, and the mastodon, which was also first named by Blumenbach, has kept the paradoxical name *Mammut*.[8]

With a better knowledge of fossil forms, the classification of proboscideans became more complicated. Today the three species of Cuvier's genus *Elephas* have become three distinct genera (*Loxodon*, *Elephas*, and *Mammuthus*), each with a separate history. The woolly mammoth is only one of the terminal branches of the extinct genus *Mammuthus*,[9] which has its own history independent of the other Elephantidae. In the course of the nineteenth and twentieth centuries, the work of Falconer, Gaudry, Depéret, Osborn, and Simpson led to the construction of many different evolutionary trees.

In the last few decades, however, the traditional ways of classifying extinct animals and constructing phylogenetic trees have been radically criticized. It has been suggested that these classifications, as elaborate as they are, lack rigor, for they often arbitrarily stress certain characters (usually dental characters) while others are ignored or considered to be secondary. Most paleontologists classified the species of fossil Proboscidea mainly by their dental characters, often using a single dominant trait to which other characters were subordinated. Some emphasized the number of crests on the wear surface (Falconer), the anatomy of those crests (Vacek), or the number of incisors (Cope). For Osborn, "the smallest bump on a molar has significance,"[10] and he gave it more emphasis than the characters of the skull or the skeleton.

The Taymyr mammoth at the St. Petersburg Zoological Institute, with the man who reconstructed it, Vadim Evgenievich Garutt (ca. 1950). (V. E. Garutt personal collection.)

Moreover, the relationships between species were often inferred from stratigraphic sequences rather than from the characters themselves: the principle "all that is ancient is primitive" seems implicitly to govern these classifications. In certain cases the ancestry is supposed to be known, which suggests that the ancestral species has been completely studied, whereas in other cases it is only hypothetical, and the ancestral groups are only defined by negative evidence—by the absence of characters found in their descendants. Genealogical trees are supposed to show the filiations and the stratigraphical succession of species, and even the processes by which they evolve. These treelike schemes also aim to describe certain presumed modes in the evolutionary process, such as speciation by geographic separation or "adaptive radiation."

In trying to cram too much information into a single diagram containing many implicit a priori, these traditional phylogenetic classifications achieved only vagueness. For this reason, some naturalists in recent decades have used methods that take advantage of the computerized coding and manipulation of data to suggest new modes of classification that would account for the totality of known characters, in order to build phylogenetic trees. Some of these new systematists saw in this formalization of classifications a way of making paleontology truly "scientific," by eliminating the need for a narrative discourse.

To classify is to count, and not to tell. This could be the motto of these "pheneticists," who essentially proceed by statistical accounting. Launched in the 1950s as a reaction to traditional methods of classification that were judged too "impressionistic," numerical taxonomy (or "phenetics") rejected any a priori evolutionary considerations.[11] Pheneticists based their work on the statistical study of characters and tried to establish "global similarities" among species by a quantified approach that rejects all prior theory, that is, all assumptions about evolutionary processes. As it happens, the very nature of the objects studied by paleontology—bones and teeth—lend themselves particularly well to a computer-based approach.

By limiting itself to descriptions, calculations, and statistical accountings of characters, numerical taxonomy claims to turn paleontological classification into an approach that is at last truly scientific, being "quantified." It proceeds by evaluating the affinities and distances between species based on the number of characters they share. But can numerical taxonomy claim to guarantee the "scientific" status of paleontology? Is there not something of a positivist illusion in this will to abolish the narrative, and even living beings themselves, in favor of lists of characters? Paleontology would then become a paradoxical science of evolution, which would not represent the parameter of time in its classifications and would consider only characters, and not organisms. In order to compare species from an evolutionary point of view, pheneticists introduced the notion of a "global resemblance," which most paleontologists find hard to accept.

Other taxonomists (the cladists) think on the contrary that traditional phylogenetic classifications, however formalized they may be, are inevitably the basis of a story in the shape of an evolutionary "scenario." Their method aims both to formalize the classifications and to situate characters and named species within an evolutionary context. This new taxonomic method was founded in 1950 by the German entomologist Willi Hennig[12] and began to spread through paleontological circles in 1966. Unlike numerical taxonomy, cladistics aims to construct phylogenies not of characters but of organisms. In its classifications, it takes into account filiations and, to a certain extent, time.

Cladists recognize that it is impossible to represent all parameters in a rigorous classification. While they consider it necessary to formalize their methods, based on a statistical study of characters, they also try to take into account the descent of species with modification through time. But in this perspective, they agree that the evolutionary process itself escapes the systematist. Neither environmental factors, natural selection, nor genetic processes can be directly represented in a phylogenetic tree. One cannot claim to represent in it evolutionary "scenarios" or the causes of evolution, whether their components are biological, ecological, or otherwise.

Cladists describe evolution by focusing not on its *processes* but on its *products*. They formalize classifications and define binary branchings ("sister groups"), not "common ancestors." Two taxonomic groups (A and B) are defined as "sister groups" if they are related to each other more closely than to any other taxon (C). The proof of the phylogenetic relationship between A and B is the presence of characters that they, and they alone, share. In this way, hypothetical ancestral groups defined without rigor are eliminated. The proximity of groups is strictly defined by their commonality of characters, which is established by counting the distinctive traits they share that result from common ancestry. This approach tends as much as possible to eliminate approximations and fuzziness from the paleontological method and discourse. In contrast to classifications based on single weighted, or "key," characters, the cladistic approach tries to use *all* the pertinent variable characters without establishing a hierarchy among them. Based on the study of their polarity, successive appearance, and shared derived characters (called "synapomorphies"), cladistics constructs a classification with minimal reference to evolutionary processes.

In a 1990 article, the French paleontologist Pascal Tassy proposed a general classification of proboscideans based on "the cladistic analysis of 136 morphological characters,"[13] in which he tried to integrate all of the variable characters formerly noted by paleontologists. In this paper, the recapitulation of previous classifications does not serve as a show of erudition or a way of rejecting earlier concepts. Instead, it presents the different parameters that cladists must take into account in developing their classifications. The classification produced according to cladistic methodological principles appears as a synthesis of all earlier morphological descriptions. Tassy writes: "This type of analysis is in fact the logical result of all of the attempts proposed for the last century and a half. We have seen that all these efforts were founded on one character or on a hierarchy of characters. Today, parsimony analysis allows us to deal with the greatest possible number of characters."[14]

Computer-assisted analysis, made possible thanks to specially developed software,[15] allows one to handle a large number of characters and organize the living beings that share them into trees that give a picture of genealogical succession. This classification thus allows one to "give an unequivocal definition of the major taxa that constitute the Proboscidea" and to define the twenty-two terminal taxa included in this group while establishing their degree of kinship.

The cladogram still resembles a tree, but its dichotomous branchings no longer purport to represent the totality of evolutionary parameters, only the

```
                         1        2        3        4        5        6      6
             0           0        0        0        0        0        0      7
Sirenia           000000000??0100??11000??????00000000000000000000000000000000
Desmostylia       000110001000100000000001000010000000000000??00000000000000000
anthracobunidés   ??????????????????????????????????????????????0?000???????????????0
Numidotherium     11?11111110101011001?001100??000000100000?000??000000?0000010011100
Barytherium       ?????????10111??1???1?0???0?1?????????????????0?000???????????1?00
Moeritherium      11111111110010001?0000001?00000000000000010000?00000000000001100000
Deinotheriidae    1111111111111111111111111111100000001000?0000?200020000000000000100
Palaeomastodon    1?111??1???111?1?????11?1111??11?????????????????102000?0?0?????00?00
Phiomia           11111111111111?11111111?111??11121110000000000?101000?0000000000000
Hemimastodon      ?????????????????????????????????????????????????????????????????????
Mammutidae        111111111111111111111111111111112211111110000?01010000000000000000
Amebelodontidae   111111111111111111111111111111110221111111211?11110000000000000000
Choerolophodon    1111111111?1111111111111111?1022111111111111?111000?000000000011
gomphothères1     11111111111111111111111111111110221121111211111110000000000000000
gomphothères2     11111111111111111111111111111110221121111211?111000000000000000
Tetralophodon     1111111111111111111111111111110221112111111211??1111000000000000000
Anancus           11111111111111111111111111111111002211121111121111011000000010000000
Paratetralophodon 111111111?1??????1111???????00211112111112?????1?????00001000?00?
Stegolophodon     111111111?1??????1111???????00221121211111?2????01?01????00110000000
Stegodon          11111111111111111111111111111100221121211112?1??01101000000010000000
Stegotetrabelodon 111111?11?111111??11?111?1?00?1?????????1?????1120??????10?00?00
Primelephas       ?????????????????????????????????????????????????0?11????????1?????0
Stegodibelodon    ????????????????????????????????????????????????01101????????????0??0
Loxodonta         11111111111111111111111111111100201112111121110111011001000010000000
Elephas,Mammuthus 11111111111111111111111111111100201112111121110111111111111110000000
```

Proboscidean character matrix by Pascal Tassy, "Phylogénie et classification des *Proboscidaea*," *Annales de paléontologie* (Masson, 1990).

relationships of species and the logic of their succession. Just as Jorge Luis Borges's short story "The Garden of Forking Paths"[16] illustrates the possibilities of narrative, the image of the cladogram embodies the very mechanism of the evolution of species, which operates by dichotomous separation. The branches of a cladistic diagram divide two by two. The "tree" with its bifurcating branches describes evolution; it concretizes in a spatial image the shape of a narrative, the successive genesis of living organisms in the course of evolution. What matter are the "leaves" (the terminal taxa, differentiated two by two), which are named as "groups." Straight, not curved, branches designate logical relations and not just vague derivations.

To the evolutionary tree thus constituted is added—or is superimposed—a text that precisely describes the successive "nodes" of the tree where the dichotomies arise, that is, the characteristic morphological traits shared by the groups that are thus differentiated. This text, which is descriptive and enumerative rather than narrative, lists the characters that distinguish different species from one another. Their genealogical succession implicitly suggests the dimension of time, although the genealogical tree does not represent the duration of time, but only morphological "distances" and evolutionary "steps." In this way, the cladistic method accounts for the dimension of evolution in paleontology as an indispensable parameter in reconstructing the history of life.

In the cladogram, the metaphor of the genealogical tree is preserved. It may be a very stylized tree, but it has "nodes," "leaves," and "roots." It does not really have a trunk, however, for there are only "sister groups" and there is no attempt to reconstruct the "missing link" or "common ancestor."

The construction of these trees is based on a few minimal evolutionary principles. The fundamental principle of this classification is that of parsimony, or the law of economy of hypotheses, which holds that generally—except in the rare case of "convergence"—a character does not appear independently twice during evolutionary history. "From a methodological point of view, parsimony analysis gives the only phylogenetic schema that can be tested on a simple logical basis: the maximization of synapomorphies," writes Tassy. "This logical basis is simply the application of the transformational model of descent with modification. This model in fact requires us to assume that most modifications are due to descent."[17]

Classifications must therefore take into account that anatomical characters common to two species have been transmitted to these descendants from a common ancestor. But this construction does not mean that certain traits cannot appear independently in certain forms. In that case, one must speak of "convergences." For example, the lower tusks that are present in several groups of Proboscidea have been lost five times in the course of their history.

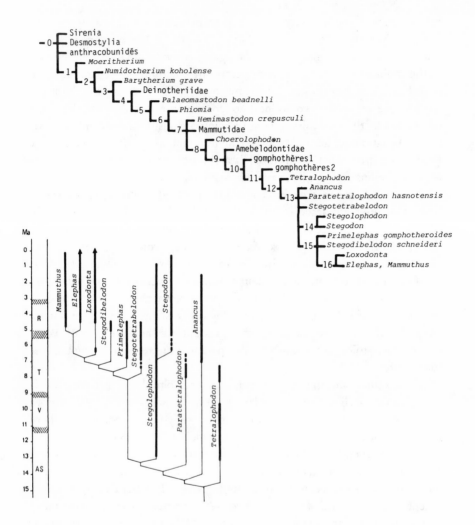

Cladogram and "parsimony analysis" tree based on the proboscidean character matrix, by Pascal Tassy, *Annales de paléontologie* (Masson, 1990).

But the occurrence of convergences is rare enough not to invalidate the method as a whole.

The history of the proboscideans in all its complexity appears to be an excellent example to illustrate the effectiveness of this method of classification, which is rigorous because it is open to refutation and can take into account the eventual discovery of new characters.

Tassy's conclusions from his study of the proboscidean family are expressed in a cladogram with sixteen "knots" corresponding to the successive differentiation of genera, the last one being the separation of *Loxodonta* on the one side, *Elephas* and *Mammuthus* on the other. From this schema it appears that mammoths share more characters with Asian than with African elephants. But "a cladogram," Tassy concludes, "does not claim to represent *the* phylogeny, any more than traditional trees do. It is but the objective (in any case refutable) expression of the basic data: the characters."[18]

A more limited cladistic study carried out by a team of American biologists led by Jeheskel Shoshani allowed a detailed reconstruction of the tree of the three genera *Mammuthus*, *Elephas*, and *Loxodonta*.[19] It identified, writes Shoshani, "six synapomorphic or shared-derived characters to the *Mammuthus* branch, thirty-eight to the combined *Elephas-Mammuthus* branch and twenty-eight distributed between the *Mammuthus* and *Loxodonta* lineages. Calculated evolutionary rates show that among the Elephantidae genera studied, *Loxodonta* is the most conservative; the monophyletic branch of *Mammuthus* and *Elephas* evolved 1.6 times faster than *Loxodonta*."[20]

Cladistic analysis thus tends to confirm the proximity of the mammoth to the Indian elephant. These results lend support to the classical hypothesis based on the study of a few traits of the skeleton and teeth. In turn they are also tools that will make it possible to identify unknown bones in the field or in museum collections.

This process of classification is performed in three successive phases. The *cladogram* represents the relationships between species taken two by two and their genealogical connection in a logical schema; the *tree* inscribes the cladogram into a stratigraphic succession and reinserts the parameter of time in the classification thus established; the *scenario* interprets the genealogical construct in terms of biological and environmental processes. So, far from being the first step and preceding classification, the scenario is its end product. It is how the analysis of evolutionary, ecological, and biogeographical mechanisms is finally introduced.

First used in zoology in 1966, cladistic analysis has been much more broadly applied in paleontological work since computers became widespread after 1980. David Hull has studied the impact of this new method on the scientific world and the often violent arguments that surrounded it, as well as all the scandals it provoked.[21] Perhaps because cladistics went against traditional representations, it has been judged to be frivolous or even dangerous. Its logical system of classification broke once and for all with finalism, teleology, and vitalism. The cladist's tree does not describe evolutionary trends; it merely expresses the description of characters. It certainly breaks with the traditional method, which *begins* with the evolutionary scenario. It is no longer a matter of reconstructing a hypothetical common ancestor—even by negative evidence—but of representing the branchings of the tree, the very products of evolution.

To define the relationship between organisms, to statistically study the distribution of characters, to quantify science—all these are ways of getting away from traditional classifications. The results of these classifications are sometimes quite unexpected. Cladistics destroys the great traditional classes (fishes, birds, mammals, reptiles) and puts birds and crocodiles with dinosaurs, for example, because they come from the same ancestral species, in a single group called Archosauria. In the past, these positions have been seen as scandalous. At the British Museum, for example, the arrangement of the dinosaur exhibits using cladistic classifications caused a noisy scandal and lengthy polemics in the columns of the prestigious English journal *Nature* in 1980 and 1981. Yet many biologists now see cladistics—phylogenetic systematics—as a tool that helps them better understand the evolution of biological diversity. From this point of view, the approach to and classification of fossils does not break with the descriptive and morphological traditions in paleontology; to the contrary, they make them necessary. Despite quantification and the use of computers, the cladistic method applied to fossils requires a study of the morphology and mineralized parts of the skeleton and teeth. It aims primarily at making optimal use of paleontological material by synthesizing all the described morphological traits, arranging them in a classification that is both logical and phylogenetic. Though formalized, it does not eliminate the narrative; it encourages the development of scenarios, which are explanatory hypotheses. "The main utility of scenarios," writes Eldredge, "is that they are capable of giving us ideas—lower-level hypotheses we might very well be able to test . . . which force us to stretch our imaginations. . . . Thus, as a heuristic device, scenarios seem useful. They are certainly more fun to construct than a mere tree or a dry cladogram."[22] They also translate the necessary persistence of *fiction* into the most formalized expressions of paleontological knowledge.

Within less than two decades, new possibilities of studying the phylogenetic relationships among organisms have opened up thanks to the techniques of molecular biology. This laboratory research has profoundly changed our approach to the history of life. It is now possible to know the processes of transformation at the level of the genome of living species, to evaluate distances and proximities among species by studying their DNA and the albumin and collagen contained in their bones, and even to determine the moment in history when they diverged. In this way molecular biology may allow us not only to study the evolutionary mechanisms of living organisms, but also to try to specify the evolutionary relationships among them. These techniques make it possible to read DNA sequences, that is, to decipher the genetic message, which consists of an arrangement of four chemical substances (nucleotides) in infinitely varied combinations.[23]

By reading the information inscribed in the nucleus of cells from living organisms, and by comparing DNA sequences between closely related species and quantifying their differences, biologists can reconstitute what they call a "molecular phylogeny." André Adoutte explains: "All the information about an individual organism as a member of a species, genus, or higher taxonomic group is coded in its DNA, and that is where evolutionary change is recorded. So relationships among species and the genetic processes by which they evolve can be studied by directly comparing the genetic structure and the organism."[24]

What is involved is no longer the presentation of divergences at the anatomical and morphological (*macroscopic*) level, as in traditional paleontology, but at the *microscopic* level, by studying "the accumulation of genetic differences at the heart of cells" and by comparing homologous structures. These techniques of comparing DNA sequences are made possible by the fact that genes have been conserved, in some cases throughout the whole history of life, that is, for 4 billion years.[25] They allow molecular reconstructions of evolution at the immense scale of the history of the living world, but also at smaller scales, down to that of a single genus, or even a single species. So by basing the approach on genetic distances (that is, by counting the differences in nucleotides between species, taken two by two), one can reconstitute a network linking species to each other and use this to build a tree representing phylogenetic evolution.

Until recently, biochemical techniques were applied only to the study of living organisms. But since the 1980s a number of biologists have tried to apply them to fossil organisms, convinced that traces of proteins may have survived in very ancient fossils. Once again mammoths became objects of special study. In a way that is unique in the history of the fossil animal world, they present molecular biology with rare and fascinating material. Their frozen or dried remains made it possible to collect certain data that until then could only be obtained from living animals. With the mammoth, it was possible for the first time to carry out a molecular study of the phylogeny of an extinct animal, just as it is done for living organisms. In this case one can really speak of paleomolecular biology.

Discoveries of frozen mammoths—at least those that have been examined scientifically—are exceptional, but they do occur. Adams's mammoth, discovered in 1799 near the Lena River estuary, was originally whole, but the scientists from St. Petersburg arrived too late, and the Zoological Institute was only able to save the skeleton. The scientific expedition that brought the

Berezovka mammoth back in 1901 managed to preserve the soft parts, frozen and packed in straw. The result was three volumes of studies on the animal's remains. Herz's team was able to perform microscopic studies of the mammoth's tissues, hairs, and even its feeding habits, since chewed plant material was found preserved in the animal's mouth. But the soft parts had been thawed and refrozen several times during excavation and transportation, and molecular studies on these remains would be difficult today.

More refined methods of preserving remains have been developed, and new specimens have been discovered, ranging in age from 10,000 years old (Yuribei) to 40,000 years (Magadan) and 53,000 years old (Khatanga). In 1977 an almost complete mammoth calf was discovered near the Dim, a small branch of the Berelyokh River in the Magadan Peninsula in extreme eastern Siberia. One Alexis Rogatchev was leveling a pile of gold-mine tailings when he saw a strange animal in his bulldozer's blade. The body was flexible and covered with hair, but completely emaciated and without any body fat. It looked as if it had died only recently, but was in fact the body of a baby mammoth, six to ten months old at death, that had been preserved in the permafrost for nearly forty thousand years. It was about three feet long, weighed just over 220 pounds, and still had all its organs, skin, and hair. The body was intact except for the right flank, which was unfortunately torn by the bulldozer. St. Petersburg Zoological Institute's Nikolai Vereshchagin, the greatest Russian mammoth specialist, rushed to the site with a team of paleontologists and began to study this extraordinary specimen.

Why were the scientists so excited by "Dima"? Because this was the first time that they had an entire mammoth at their disposal. And its juvenile characteristics made it especially interesting, because they yielded invaluable information on the growth of mammoths, which was poorly known. Dima also presented its discoverers with many enigmas. What circumstances led to its death? Several scenarios have been suggested. In one, the baby mammoth escaped its mother's vigilance, fell into a hole too deep for it to climb out of, and died trapped and exhausted. This would account for its extreme thinness and the total absence of subcutaneous fat when it died. Others noted that the little mammoth had a wound on its right leg, probably caused by a pointed object shortly before the animal died. Had Dima and its mother been tracked and wounded by hunters? It is a plausible hypothesis, though no evidence of human activity from those times has been found in the area.

Dima's organs and tissues, as well as its geological and vegetable environment, have been studied in great detail and have produced extensive publications.[26] Today, however, that tissue cannot be studied at the molecular level because shortly after the find, its body was preserved by being soaked in a

"Dima," an eight-month-old baby mammoth found in Magadan (eastern Siberia) on June 23, 1977. It weighed ninety kilos and had been buried there for thirty-nine thousand years. The first studies on mammoth DNA were performed with the flesh from one of its thighs. (V. E. Garutt personal collection.)

vat of paraffin. Its hair fell out during the operation, and all that is left of the little mammoth is a black, rubbery carcass. But before this "embalming" procedure, Russian scientists took some frozen muscle samples, which were sent in 1978 to Berkeley, California, where Dima became the subject of some of the earliest studies in paleomolecular biology.

In 1981 Jerold Lowenstein tested the albumin from Dima's right thigh muscle, to try to measure the degrees of relationship between the mammoth and the Indian elephant, on the one hand, and the African elephant, on the other.[27] In the laboratory, biologists set about "cooking" to extract these precious substances (genetic material) from the frozen or dried tissues of these extinct mammals to compare them to those of their living cousins. Lowenstein describes the way they put them through grinders and centrifuges to extract the proteins, chopped up the flesh and the bones, marinated them in salt and chemical substances, bombarded them with radioactive beams, exposed them to different temperatures, and photographed them with electron microscopes. Lowenstein's study made it possible to show that mammoths are related to the elephants of both Africa and India, and that these three branches diverged between 3 and 5 million years ago, which matches the paleontological data. In that way, molecular systematics of an extinct species was achieved for the very first time. The proteins contained in the mammoth tissues retained enough of their original structure so that their phylogenetic relation to other species of living elephants could be established. The body of these immunochemical studies also shows the excellent preservation of *Mammuthus* tissue at both the structural and chemical level.

Will the study of affinities based on mammoth DNA and its comparison with those of living elephants allow us to reconstruct more finely the evolutionary history of the Elephantidae? Is it possible to determine with precision whether the mammoth is closer to the Asian or the African elephant? For a long time, scientists had to be satisfied with very general conclusions. The California team that first attempted to extract mammoth DNA and compare it with that of other living proboscideans concludes:

> Phylogenetic relationships based on immunochemical results obtained thus far are equivocal, with a slight indication of the traditional hypothesis. This may be interpreted as indicating a slow rate of molecular evolution, since paleontological evidence shows that members of the Elephantidae diverged at about 4.5 million years ago. An effort is being made to establish a sound phylogenetic relation within the Elephantidae based on the materials available. This is extremely difficult, because we are dealing with protein molecules (mostly collagen and albumin) that are evolutionarily conservative and, in addition, are probably not preserved in their native state.[28]

Indeed, studies based on proteins at the time were not refined enough to confirm the conclusions of morphological studies, namely that the mammoth is closer to the Asian elephant, with the two species having separated more recently.

New studies have been conducted on mammoth DNA, mainly in Japan, Russia, the United States, and France. In 1999, using techniques developed by a Russo-Japanese team[29] and working mainly on mitochondrial DNA, Régis Debruyne, a doctoral student at the Paris Muséum National d'Histoire Naturelle, performed some remarkable experiments on the remains of a mammoth from the Lyakhov Islands now kept at the Paris gallery of paleontology. This mammoth was found in 1910 by the Russian paleontologist K. A. Volosovich. Deeply in debt, he borrowed money from his rich friend Count Stenbocq-Fermor but was unable to pay him back and gave him the mammoth instead. The count brought parts of the mammoth (the skeleton and some of its soft tissues, mainly the feet) to Paris and presented it to the French government in hopes of being awarded the Legion of Honor. The story does not say whether Stenbocq-Fermor got his medal, but almost a hundred years later, molecular studies could be carried out on the remains of his mammoth. Debruyne first tried to extract DNA from the skin, but without any significant result. He eventually succeeded in extracting, amplifying, and sequencing DNA from a bone of the tarsus and came to the unexpected—not to say revolutionary—conclusion that the mammoth may be phylogenetically closer to the African than to the Asian elephant.[30]

Thus, extraction, purification, and sequencing of ancient DNA yield both puzzling and interesting results, as they seem to contradict classifications previously established by paleontology. This raises the problem of the convergence between paleontological phylogenies and taxonomies achieved by molecular study. When anatomists and biologists compare their results, in the best case they confirm each other. But there are often disagreements, in particular in the estimate of *time*, the duration of the evolutionary process. Paleontologists deal with geological "deep time," and they can look for "concrete" evidence of evolution in the stratigraphic succession of fossil remains, while biologists ponder the "molecular clock," the rates and rhythms of evolution. Divergences between molecularists and paleontologists may also exist in their very conclusions concerning the history of species. In this case as in many others, only further studies and close collaboration between specialists in the two disciplines will help to solve the puzzles of the evolution of life.

Paleomolecular studies also have a bright future because they suggest the possibility of cloning fossil DNA, which would allow the sequencing of mammoth DNA and direct comparisons with DNA from other mammals.

> The most promising aspect of future work on the biochemical analysis of mammoth tissue is the application of recombinant DNA and other molecular genetic techniques to the cloning

and analysis of mammoth DNA sequences. . . . The results should permit the opportunity for studying directly, for the first time, the gene structure of an extinct mammal and offer the possibility of investigating a number of important fundamental questions in evolutionary biology.[31]

Cloning mammoth DNA would prove that genetic material can survive more than fifty thousand years. This would extend our knowledge of the DNA of extinct animals very far back in time. It would also open the door to all sorts of dreams and fantasies. Since the most publicized success in cloning the sheep Dolly, the idea of cloning a *whole mammoth* by inserting mammoth DNA into the enucleated egg cell of a female elephant has become a very widespread dream.

When it was thought that a complete frozen mammoth was found in the Siberian tundra in 1997, the main motivation of those who decided to save the carcass was to clone a mammoth. The Dolgans, nomadic natives of the region, had alerted the authorities in Khatanga, on the Taymyr Peninsula in northern Siberia, to the presence of a frozen mammoth. French adventurer and aviation buff Bernard Buigues, who has organized polar tourist expeditions, soon got interested in the affair and became fascinated with the mammoth.

At his own expense, Buigues gathered a team of amateurs, technicians, and a few French, Russian, and American scientists to organize an expedition to extract the mammoth dubbed "Jarkov" from its icy sarcophagus. Buigues and his friends said that they hoped to extract "genetic material in good condition" from the animal, whose carcass had been frozen in the permafrost for 20,300 years. This could soon result, they said, in "baby mammoths gamboling on the tundra or in the Bois de Vincennes" in Paris. After all, had not one of his scientific advisers claimed to have discerned with an electron microscope a cell wall in a bone fragment that Buigues brought back from Siberia?

It was eventually decided that the animal should be preserved whole in the ice, so as not to interrupt the "chain of cold," thus preserving its organs and genetic material intact. This did not stop the team members from thawing the hairs with a hair dryer, which would make any of the precious genetic material that might remain in the hair follicles permanently unsuitable for molecular study.

The specimen was carried in its block of ice, to one of the Khatanga caves that Stalin had ordered dug during the cold war as a nuclear bomb shelter for the population. Samples of its tissue have already been sent to various laboratories around the world, but talk of cloning has stopped, and it seems unlikely that it could ever be carried out. Up to now, not a single intact

Jarkov the mammoth found on the Taymyr Peninsula, Siberia.

mammoth cell, however well frozen, has ever been found. Biologists have only been able to extract only small samples of very degraded fossil DNA, even from animals that seem to have been very well preserved after death. As for decoding the mammoth genome—another ambition of the team gathered around the Jarkov carcass—that seems to be a unlikely dream, since it would require huge quantities of DNA. "At best, one might hope to sequence the mammoth's mitochondrial genome (about sixteen thousand nucleotides)," says Régis Debruyne. "As for its usefulness, that's hard to say. It would be useful for fundamental research into the history of the lineage, but probably not of much use to the person on the street."[32]

What remains of the expedition that set out to find Jarkov during 1999 are the luminous images of these northern lands, which appeared in newspapers and on television around the world. The aviator Buigues may not have succeeded in cloning his mammoth, but he was at last able to fulfill his dream of making it fly above the Siberian tundra, trapped in its icy sheath, carried by helicopter at the end of a cable.[33]

Other serious projects have been formed in order to "resuscitate" the mammoth. Since 1990 Kazufumi Goto, a Japanese professor of animal reproduction at the University of Kagoshima, has dreamed of resuscitating woolly mammoths by injecting some of their sperm into the ovocyte of a female elephant to produce baby "mammoth-elephant" hybrids. If after selecting the newborn females and letting them grow, you repeat the operation, you might eventually get an animal that looks much more like a mammoth than an elephant! In Japan an association named Re-creating the Mammoth has been specially set up to fund this project, and many publications, including *New Scientist*, have popularized Goto's idea. But his ambitious plan has received a generally lukewarm reception in the scientific world. Inseminating female elephants has never been achieved successfully, as their uterus is more than four feet long and is curved in a way that makes the operation very delicate. And until now, no mammoth sperm has been found well preserved enough to make such an experiment possible. It is most unlikely that spermatozoa would remain in good shape or that a whole cell with its DNA intact would survive after spending several thousands of years in the ice.

Nevertheless, Goto is pursuing his project of "hunting mammoths" in Siberia in order to find a frozen animal with testicles containing intact spermatozoa. High-tech expeditions that he organized in Yakutia—the "motherland of mammoths"—have come to a dead end. Is this an absurd dream, or does it require only time in order to succeed? After all, men have gone to the moon, and the technology of assisted mammal reproduction is progressing rapidly. Today several teams in the world are working hard at extracting and studying fossil DNA. "Can we hope to see . . . mammoths stalking the earth again?" asks English paleontologist Michael Benton. "Molecular biologists merely have to develop techniques to insert the cloned DNA of . . . the mammoth into the early embryos of developing . . . elephants, and who knows what might happen?"[34] In any case, this dream is more alive today than ever. In June 1999 a group of eminent paleontologists organized a memorial service at the Mammoth Site in South Dakota to mourn the extinction of the mammoth and expressed the wish to revive lost Ice Age megafauna in the American West by reintroducing its closest living relatives.[35]

So once again the mammoth finds itself at the cutting edge of the most advanced techniques and the wildest fantasies about reconstructing the past. The possibility of cloning and sequencing mammoth DNA revives the old dream of seeing extinct animals live again, the alchemical theme of the phoenix reborn of its ashes, the religious myth of the resurrection of the dead. By giving life to extinct species, paleontology would surpass itself and move from being a science of death to truly being a science of life in every sense of the word. It would rediscover the circular image of eternal time, in which mammoths would gaze at us throughout eternity, at once worried and thoughtful, their deep eyes lost in the reddish mass of their hair.

Life and Death of Mammoths:
Scenarios for an Extinction

As the blizzard rages, the mammoth keeps its young close. In the dark, storm-whitened wool of its flanks, one can make out a small ball of black fur, rigid with cold. In the distance, the herd is scattered and disorganized. The image seems blurry. The wind is howling across the frozen wastes; the snow, falling in large, wet flakes, will soon bury them.

A fragile giant, the mammoth is doomed to die.

Many paintings by the Czech illustrator Zdenek Burian portray this end.[1] *The Mammoth and Its Young in the Blizzard* (1961) illustrates one of the "scenarios" developed at the beginning of the century by the American naturalist George Wright: "As the climate gradually became more and more severe and the summers shorter and shorter, the inertia of this migratory spirit continued, and large herds of mammoth from time to time were caught in the fearful blizzards, so common now during the early autumn in northern Siberia, and perished from cold and hunger."[2] Other paintings suggest different scenarios. Hunger: the mammoth searching the earth for its meager nourishment, breaking the surface of the frozen ground with its curved tusks. Accidents: the dying Berezovka mammoth trapped in a crevasse from which it cannot escape. Humans: men tracking the herd, driving mammoths toward traps and butchering them.

In Burian's well-known illustrations for the books by paleontologist Josef Augusta,[3] the artist showed the menace threatening this colossus of the ice in various ways. Perhaps because of the extraordinary quantities of mammoth bones discovered in the Paleolithic dwelling sites in Moravia, which suggest a massive killing of these animals, perhaps because of the more somber postwar political climate in eastern Europe, Burian's mammoths do not have the triumphant, conquering appearance that Charles Knight painted in the United States several decades earlier. At midcentury extinction was the subject of

The Mammoth and Its Young in the Blizzard (1961), by Zdenek
Burian. This painting illustrates one of the "scenarios" developed at
the beginning of the twentieth century to explain the extinction of the
mammoth. (Moravské Zemské Muzeum–Anthropos, Brno.)

obsessive interest. The mammoths have disappeared but their bones remain,
sometimes in such huge quantities, on the great plains of central Europe and
on the coasts and islands of northern Siberia, that the ground is literally covered with them. And their frozen bodies, which are sometimes discovered
in the most northern parts, make their death even more present and more
dramatic.

The extinction of the mammoth some ten thousand years ago—like the
extinction of the dinosaurs 65 million years ago—remains in many respects a
scientific enigma. Constructs and narratives suggesting probable explanations
have been built to account for it. But probability itself has a history, and a
number of hypotheses that were accepted in the nineteenth century to resolve
the problem are under attack today. Is the extinction of species a consequence
of changes of the environment, whether by cataclysms or slow transformations of the milieu? Is it a result of a destiny, of "laws" of life that prescribe

the birth, life, and death of a species? Is it linked to the struggle among species for their survival, or to the destructive activity of man? These different hypotheses, associated with varying representations of time and of causes at work in nature, suggest multiple narratives, based on different systems of thought and ways of representing the world and man.

Today the question of extinctions is one of the most heated, and the most popular, in paleontology. It is connected with our preoccupation with endangered species and with the ecological balance of our planet. For this reason it carries a passionate, emotional, and even mythical charge. What is involved is death—*our* death—the possible or probable extinction of our species. Though less spectacular than that of the dinosaurs, the extinction of mammoths is closer to us and more familiar. It is also more meaningful, since it relates to the destiny of the human species.

How do species die? Why did the mammoths disappear?

The question of extinctions was the first scientific inquiry of paleontology. As early as 1580 in his *Discours admirable des eaux et des fontaines* (An admirable discourse on waters and fountains), Bernard Palissy spoke of "lost species" and hypothesized that the unknown fossil fishes and shells he collected in Saintonge were the remains of animals that became extinct because man had overfished them.[4] A century later Leibniz was still wondering about the fossil shells of unknown animals—of which the most famous at that time were ammonites, or "Ammon horns." But Leibniz believed that God had created a perfect and immutable world, so the disappearance of species remained unthinkable for him.[5] Buffon accepted that the cooling of the climate could cause certain species to disappear but felt they "degenerated" rather than became extinct and that their "inner mould" survived in another form.[6] At the end of the eighteenth century, Cuvier placed the question of "lost species" at the center of his narrative construction and scientific system.[7] In the extinction of the great antediluvian pachyderms, Cuvier saw the effect of huge cataclysms that had annihilated successive fauna. The fact that species were mortal, and that the animals whose remains were found in the successive layers of the earth had become irreversibly extinct, made it possible to put forward the very notion of a history of the earth and to order geological events in a chronological sequence. The geologist Alcide d'Orbigny defined twenty-seven successive geological "stages" according to the species of fossil invertebrates identified within them.[8] A few decades later he showed that knowledge of extinct species allowed one to precisely date the strata that contained them.

While claiming that entire fauna had been annihilated by gigantic "cataclysms," Cuvier did not precisely identify the exact nature of these catastrophic events. Were these brutal climatic changes, sudden cold snaps, as suggested by the presence in the permafrost of the carcasses of Siberian mammoths that had been caught unawares and froze to death? Were these movements of water, violent tidal waves, or, for that matter, "deluges"? Cuvier did not really answer the question. In England around 1820, the geologist William Buckland and his disciples, who believed in natural theology, had firmly accepted the hypothesis of a diluvial cataclysm.[9] This is why Quaternary soils were long thereafter known as the "*diluvium*." But the attempts to make the history of the earth coincide with the episodes of the biblical account were gradually abandoned. Miracles and divine punishments were no longer enough to account for the history of the world. In 1887, when Henry Howorth was still defending the theory of a diluvial catastrophe to explain the disappearance of mammoths in a book entitled *The Mammoth and the Flood*,[10] his position seemed long out-of-date. New theories had been put forth in the meantime.

The Swiss naturalist Louis Agassiz—a fervent disciple of Cuvier who had begun his work on fossil fishes at the Paris Muséum National d'Histoire Naturelle in 1832—maintained the doctrine of catastrophes until his death in 1873. But Agassiz abandoned the traditional belief in "deluges" for a new explanation: the theory of glaciers. Switzerland, where Agassiz spent the first part of his life before emigrating to the United States in 1847, is a particularly favorable place for observing glacial phenomena. During the 1830s two naturalists named Venetz and Charpentier, who were members of the Swiss Society of Natural Sciences, noticed "large erratic blocks" and great moraines—exactly like those that lined glaciers in the lower valleys of the Alps and on the slopes of the Jura Mountains—lying at some distance from contemporary glaciers. These formations suggested that in the past, glaciers there stretched all the way to the shores of Lake Geneva.[11] From the presence of these moraines and erratic blocks in places sometimes very far from modern glaciers, Agassiz (who was the young president of that honorable scientific society) concluded that glaciers must have extended much farther than they do today, "from the North Pole right down to the edges of the Mediterranean and Caspian Sea." Glacial valleys in Switzerland and throughout Europe and North America still bear their traces today.

"Since I saw the glaciers I am quite of a snowy humor, and will have the whole surface of the earth covered with ice, and the whole prior creation dead by cold. In fact, I am quite satisfied that ice must be taken [included] in every complete explanation of the last changes which occurred at the surface of Europe," wrote Agassiz to William Buckland in 1838.[12]

In this way, the "sudden intense winter" that struck the earth could have been responsible for the destruction of ancient fauna, as proven by the bodies of mammoths frozen in ice in Siberia. Agassiz wrote: "Siberian winter established itself and lasted for a time over a world previously covered with a rich vegetation and peopled with large mammalia, similar to those now inhabiting the warm regions of India and Africa. Death enveloped all nature in a shroud, and the cold, having reached its highest degree, gave to this mass of ice, at the maximum of tension, the greatest possible hardness."[13]

Using this hypothesis, it is possible to construct a universal system, a new narrative of the history of the globe. Agassiz conceived of the history of the earth as the development of individual life that witnessed a hot period of life and balance, and a cold phase of destruction and death. The history of the earth is that of a slow cooling interrupted by brutal "oscillations," which represent the abrupt glacial phases. The "epochs" that Cuvier had distinguished coincide precisely with the succession of these climatic phases of warming, stable balances, and sudden cooling periods.

> The surface of Europe, adorned before by a tropical vegetation and inhabited by troops of large elephants, enormous hippopotami, and gigantic carnivores, was suddenly buried under a vast mantle of ice, covering alike plains, lakes, seas and plateaus. Upon the life and movement of a powerful creation fell the silence of death. Springs paused, rivers ceased to flow, the rays of the sun, rising upon this frozen shore (if, indeed, it was reached by them), were met only by the breath of the winter from the north and the thunders of the crevasses as they opened across the surface of this icy sea.[14]

Agassiz developed this "theory of glaciers" in accordance with the Cuvierian vision of a discontinuous history of life. The brutal glacial periods *were* the Cuvierian catastrophes that periodically destroyed all life. In 1846, upon reaching the American continent at Halifax, he noticed, "on the first undisturbed ground, after leaving the town [. . .] the familiar signs, the polished surfaces, the furrows and scratches, the *line engraving* of the glacier."[15] In 1866 in his *Geological Sketches*, Agassiz returned to the gripping and poetic evocation of this "sudden intense winter" that annihilated the fauna of giant quadrupeds at the end of the Quaternary era.

> The long summer was over. For ages a tropical climate had prevailed over a great part of the Earth, and animals whose home is now beneath the Equator roamed over the world from the far South to the very borders of the Arctics. The gigantic quadrupeds, the Mastodons, Elephants, Tigers, Lions, Hyenas,

The Zermatt glacier. Louis Agassiz, *Théorie des glaciers* (Atlas, 1840).
(Harvard Museum of Comparative Zoology photo.)

Bears, whose remains are found in Europe from its southern promontories to the northernmost limits of Siberia and Scandinavia, and in America from the Southern states to Greenland and the Melville Islands, may indeed be said to have possessed the Earth in those days. But their reign was over. A sudden intense winter, that was also to last for ages, fell upon our globe; it spread over the very countries where these tropical animals had their homes, and so suddenly did it come upon them that they were embalmed beneath masses of snow and ice, without time even for the decay which follows death.[16]

In this way the theory of glaciations was linked from the outset to the extinction of species. Throughout the nineteenth century and even up to the present, many authors have perpetuated the myth of the extinction of the mammoth by a brutal cooling of the climate. In 1961 William R. Farrand presented a careful argument in *Science* magazine to refute the idea that mammoths could have been "frozen in just a few hours."[17] In literature produced today by American believers in creation science, books and articles periodically appear stating that the mammoth's disappearance was the result of cataclysms.[18] Linked to a fixist representation of the living world, catastrophes allow for a shorter chronology to account for the history of the earth, which is more in line with the biblical time span.

It is precisely in rethinking the length of geologic times that Charles Lyell came to reject the catastrophist explanation of extinctions.[19] In the first volume of his *Principles of Geology*, published in 1830, Lyell showed that one could not invoke "catastrophic" causes, irreducible to those that we know today, as rational, scientifically conceivable phenomena.[20] The present is the key to the past. This is the axiom on which geology is founded as a science. The causes of the earth's transformation are knowable and can be formulated as laws because they are the very same ones—slow, continuous, and uniform—that are at work in the present. Their action is possible because they take place over immense stretches of time.

Having rejected catastrophes, Lyell had to propose a new explanation for extinctions, and here again the mammoth was at the heart of the debate. For defenders of "cataclysms," the Siberian mammoth played a central role. But Lyell thought that extinction was a continuous process that followed gradual changes in climate and environment, and continues at work in nature today. He was willing to accept the glacial theory in 1840 but refused to believe that the mammoths had disappeared all at once, any more than any other extinct species had. The large quantity of their remains in Siberia indicated that they survived the cooling for a very long time. "Their bones are found in icebergs and in the frozen gravel, in such abundance as could only have been supplied by many successive generations."[21]

These enormous mammals could have become extinct by the effect of "extremely slow" climatic changes that consist not in a drop in the mean annual temperature so much as a progressive change of climate during which the seasons became more and more extreme. In this way, the mammoth came to support the concept of slow and gradual extinctions. To propose a "uniformitarian"[22] explanation for the extinction of the mammoth was to challenge catastrophism on its own ground.

In the middle of the nineteenth century, a new actor, with the promise of a brilliant career ahead of him, stepped onto the extinction scenario stage: man. After 1859 the existence of *fossil man* living side by side with the great Quaternary mammals was accepted in France and England. From then on, the hypothesis that Lamarck had proposed a few decades earlier to account for the disappearance of certain species—and that Cuvier delighted in ridiculing—became plausible. Man could have destroyed "antediluvian" animals because he had been their contemporary and because he had fearsome weapons, those famous flint axes that Boucher de Perthes found in the lower Somme Valley. In the more recent periods of the Paleolithic, these instruments became more precise and perfected: the arrowheads, blades, and knives that Lartet and Christy uncovered in the Vézère Valley sites.[23] Here was a hypothesis rich in new interpretations and questions: Could man the mammoth hunter be responsible for those animals' disappearance? Toward the end of his life, Lyell himself viewed this possibility favorably, having recognized the antiquity of man and the truth of Boucher de Perthes's theories. While reaffirming that the causes of extinctions are "more general and more powerful than the action of man," he admitted in 1863 that "the growing power of man may have lent its aid as the destroying cause of many post-Pliocene [Quaternary] species."[24] In the second half of the nineteenth century, this new vision of the extinction of great Quaternary animals caused by human action was sometimes linked to theological considerations. It was developed, notably in England, by those who still believed that man is the goal of creation, destined by God to reign on earth. Which is why Samuel Haughton, a professor of geology at the University of Dublin, could suggest in 1860 that it was primitive man's role, in being placed on an earth that was still imperfect, to render nature completely modern by destroying the great mammals that still haunted it.[25]

Others of the same period imagined completely different causes to explain the extinction of species, seeing at work the effect of a destiny inherent in their organization. Richard Owen conceived of the final extinction of *Elephas primigenius* and other Quaternary species as the effect of predetermined causes, of a kind of destiny that the Divinity had fixed in advance. "It [is] almost . . . reasonable to speculate . . . on the possibility that species like individuals may have had the cause of their death inherent in their original constitution, independently of changes in the external world, and that the term of their existence, or the period of exhaustion of the prolific force, may have been ordained from the commencement of each species," wrote Owen.[26]

The disappearance of certain species must be explained through the order of nature itself. To assure balance, nature must get rid of its excess. So the mammoths would have gradually become "extinct" by the effect of their own tendencies. For each species it would be possible to define a certain length of time, a characteristic longevity in function of its own vitality. This "non-violent" extinction theory proceeds from a vitalist approach, more or less rooted in metaphysics. Like individuals, families and orders are the repositories of a certain quantity of life that cannot be exceeded. The longevity inherent in each species, as in each individual, is predetermined. The cycle of their appearance and extinction follows the natural arc of life, growth, and senescence, "aging," extinction.

At the turn of the last century, this position was held by a number of paleontologists in France and the United States. "In general, the exhaustion of the type has been as complete as its flourishing had been magnificent," wrote Albert Gaudry.[27] For the American paleontologist Henry Fairfield Osborn, the concept was linked to his desire to develop "laws" of the evolution of the living world and introduce the idea of an orthogenesis.[28]

The work of the Russian geographer I. P. Tolmachoff illustrates this view very well. Tolmachoff began his career as a paleontologist in Russia and participated in the 1901 expedition that retrieved and studied the Berezovka mammoth. He emigrated to the United States at the time of the 1917 revolution and became a disciple of Osborn, who was then the leading scientific authority in paleontology and evolutionary theory at the American Museum of Natural History in New York. The doctrine that Tolmachoff presented in his 1929 monograph, "The Carcasses of the Mammoth and Rhinoceros Found in the Frozen Ground of Siberia,"[29] was related to the context of its author's thinking. Like Osborn, he wanted to set forth laws on the evolution of life and the morphological transformations of living beings as a way to rationally account for the general phenomenon of extinction. Tolmachoff carefully examined all the hypotheses: human causes, climatic or geological causes, the separation of the islands of New Siberia, the collapse of ice along their regular routes, cooling or warming of the climate. In his eyes, the hypothesis that mammoths were poorly adapted to their environment was not tenable. Some thought that the mammoths didn't have enough to eat. But the mammoths whose bodies had been discovered were very well nourished, sometimes even too fat. Others claimed that the structure of their skin was not sufficient protection against the cold because they lacked sebaceous glands, an argument put forth by the French prehistorian Henry Neuville.[30] But the study of subcutaneous layers revealed a thick layer of fat under the skin. Moreover, the mammoth's coat, like those of other Arctic animals, consisted of two layers, a fine woolly coat and a covering of long hairs, which constitutes excellent protection against cold.

"We must explain the disappearance of an animal which was living in great numbers apparently very prosperously, over a large area to which it was well adapted," wrote Tolmachoff, "and which died out in a very short time, geologically speaking."[31] Too often, he noted, the problem has been seen under only a local or a particular angle. It was necessary instead to distinguish general principles by which extinction can be explained. Its causes must be found not in a poor adaptation to the environment but in an overdeveloped specialization. Through his notion of "hyper-specialization," Tolmachoff developed a general theory of the extinction of species, of which the woolly mammoth was a particular case. This animal, which was particularly well adapted to extreme cold, had characteristics of extreme specialization: the structure of its molars, which consisted of numerous dental plates with very thin layers of enamel; the enormous shape and spiral form of its tusks, which were practically useless in the adult; the structure of the foot, possessing four toes (instead of five, as elephants have); and the anal flap. Tolmachoff thought mammoths had exhausted their vitality by making too many successful "efforts" to adapt to particular conditions of life. The species vanished gradually for lack of "vital force," like a lamp running out of fuel. Thus, "the great achievement of an individual may become destructive for the species."[32] With his notion of "hyper-specialization," Tolmachoff borrowed from Osborn the idea that only the most "primitive" forms have the capacity to evolve and diversify in an "adaptive radiation." The more particular characteristics a lineage has, he claimed, the less it is capable of evolving and adapting to new ways of life. It can only develop according to its own tendencies pushed to the extreme, in an orthogenesis. In this way, "hyper-specialization" generally precedes the end of a species. Its loss of vitality comes from "the heroic effort by the individual to maintain the race." The rate of reproduction of mammoths fell to a very low level—a trait that is evident among elephants today—and a species could thus gradually disappear, without wars or catastrophes. In this explanation one can recognize a variation of the Lamarckian idea of the individual effort to adapt to the milieu shaped by "eugenic" theories dear to Osborn and Madison Grant, who held elitist and racist ideas about the fragility of "superior" human races and the risk of extinction they faced because of their "hyper-specialization."[33]

In the first decades of the twentieth century, theories were thus developed that gave a major role to "vital force," to the effort to adapt, and to an orthogenesis that ruled the logic of succession and extinction of beings according to general trends and "laws" of evolution.

The vitalist and teleological framework in which extinction was being considered was very far from Darwinian ideas. Extinction, for Darwin, was not the consequence of catastrophic geological causes, or laws prescribed from the outside, or a mysterious destiny of beings. The determining factors of evolution are the relationship of living organisms to their milieu, the natural selection within species, and the struggle for life and the elimination of the less fit. The struggle for survival favors individuals that are adapted best and causes those who are less well equipped to subsist in a given environment to perish. A slow and moderate pressure exercised by new species can lead others to extinction. "[A]s new forms are continually and slowly being produced . . . , numbers inevitably must become extinct."[34] Darwin used the metaphor of wedges driven into a tree trunk: if the surface of a log is covered with wedges, any new wedge driven in will expel old ones from the log.[35] The "new wedges" represented species that have recently appeared in the course of evolution. One can therefore understand why the "extinction of less favored forms" is the "inevitable consequence" of evolution. "[T]he manner in which single species and whole groups of species become extinct, accords well with the theory of natural selection," concluded Darwin.[36]

Starting in the 1930s, the New Synthesis theoreticians returned to Darwinian concepts to criticize the orthogenesis and vitalist ideas of Osborn's generation. It was absurd, they wrote, to consider extinction with a negative moral judgment, in considering, for example, that the terminal lines showed a disgraceful "senescence," "racial aging," and extinction by exhaustion of vital force. Such an approach would establish a parallel without rational foundation between the history of the individual and the history of the species.

Alfred Romer wrote: "The organs and protoplasm of an individual may age in one fashion or another, but it is obvious upon consideration at a given stage in Earth history no one type of animal is a whit older than any other. Unless animal life has come into existence more than once, each line has a pedigree just as long as every other, and should be no more senescent than any other."[37] Extinction is not determined by senescence, "hyperspecialization," or mysterious predestination, but only by the relations among living beings and their environment. A species becomes extinct because the animals encounter conditions in which they can neither successfully survive or reproduce. Lack of food may play a role, in that competition for food with other species may lead to famine for an entire species. The disappearance of animal or vegetable diet elements that a group normally eats can be a major factor in the extinction of certain species. Animals can also be decimated by illness. "All these varied factors may be summarized as environmental changes too great or too rapid for the organism concerned to adjust to them," Romer wrote.[38] In this context the notion of adaptation is no longer connected to an orthogenetic law of progress. Extinction often affects species that are of large size, abundant, and in great number—and in the case of land verte-brates, herbivores rather than carnivores. But this is not because large size is a symptom of "hyper-specialization" or an "end of the line" characteristic, but because at a certain point, size can limit an animal's capacity to adapt to new conditions. Herbivores, which are generally more numerous than carni-vores and more specialized in food gathering and in their digestive mecha-nisms, usually have more difficulty adjusting to changes in the environment.

This renewed vision largely drew on Darwin to place extinction within the framework of the evolution of life. In the *Origin of Species*, Darwin had put the question of "mass extinctions" somewhat aside, while not neglecting it completely, in that he was following Lyell's view of geological change. For Lyell, small modifications observable in the present, accumulated over thou-sands of centuries, were enough to explain evolution at the geological time scale, without having to bring in mysterious "catastrophes." Darwin ex-plained: "Species and groups of species gradually disappear, one after the other, first from one spot, then from another, and finally from the world."[39] Animals gradually become more rare, then vanish. But does our experience, which is limited to the historic scale, allow us to know *all* the processes capable of producing changes on our earth? Can one imagine at work in the history of the world phenomena whose amplitude and frequency are such that they cannot be observed by man? As Fontenelle wrote at the end of the seven-teenth century, "No rose can ever remember seeing a gardener die." If within human memory we have seen the extinction as a result of man's action of the dodo—a large bird on the island of Mauritius that was wiped out by sailors who stopped there—or the dwarf elephants of North Africa, what is our all-too-human scale to make of the causes that led to the extinction of whole fauna at the end of the Cretaceous 65 million years ago or at the end of the Pleistocene ten thousand years ago?

Since the 1980s, new attention has been paid to mass extinctions. Among certain paleontologists,[40] it is connected with the desire to revise the classic representation, inherited from Lyell and Darwin, of an evolution that proceeds gradually, eliminating the least well adapted living forms along the way. Extinction is not always oriented toward the conservation of the species that are most apt to survive in a given milieu. To the classical representation of a gradual evolution that proceeds by successively eliminating the less apt, thereby resulting in better adaptation and progress, one could oppose the vision of a more hazardous causality.

"Mass extinctions" have happened several times during the history of life, and entire fauna have been rapidly replaced on a very large scale. Among the five biggest extinctions, the most famous is the one at the boundary between the Cretaceous and the Tertiary, which annihilated the dinosaurs and a great number of animal and vegetable species, both marine and terrestrial, 65 million years ago. Compared to these "big five," the extinction that caused the mammoths to disappear at the end of the Pleistocene ten thousand years ago seems a minor event; maybe it is because it happened to our fellow mammals that we give it any importance at all. And yet there, too, it is necessary to explain the disappearance of an entire fauna—mammoth, woolly rhinoceros, cave bear, megaceros, and saber-toothed tiger. Certain species that became extinct on one continent survived on another, like the musk ox, which disappeared in Europe but not in America; or the reindeer, horse, and camel, which became extinct in America but survived in Eurasia or in Africa. "Is proneness to extinction an inherent property of species—a weakness—or does it depend only on vagaries of chance in a risk-ridden world?" asks David Raup, a specialist in statistical paleontology. Extinction could be the fact of accidental external circumstances, which are "catastrophic" for certain species. These accidental causes could have acted selectively on certain groups while having no particular adaptive direction nor any role in the process of "the struggle for survival." "Extinction is necessary for evolution, as we know it, and selective extinction that is largely blind to the fitness of the organism . . . is most likely to have dominated," Raup concludes.[41]

This explains the renewed interest in catastrophe scenarios. When Luis and Walter Alvarez suggested in 1980 that the dinosaurs were wiped out as the result of an asteroid hitting the earth, they relied on geological evidence, and their revolutionary thesis hit like a thunderclap. With such a hypothesis, are we witnessing a return to nineteenth-century "catastrophism"? Or an attempt to account for the discontinuities in the history of life against the background of a theory of evolution, opposing a representation of geological causality that had become too anthropocentric?

Today two models dominate the interpretation of the extinction of mammoths. On the one hand, environmental causes: the warming of the climate at the end of the Pleistocene. On the other, human causes: Paleolithic men in a brief time massacred entire species and thereby caused their extinction.

Supporters of the climatic explanation no longer speak of a brutal cooling of the climate as in the days of Cuvier or Agassiz. A warming period at the end of Quaternary times (between 14,000 and 10,000 years ago) and the melting of the glacial ice sheets would have caused the sea level to rise and glacial lakes to dry out, causing ecological perturbations that threatened large mammals. Abrupt alternations between hot and cold at the passage from the Pliocene to the Holocene changed the environment into a swampy tundra, conditions to which the mammoths could not adapt. In northern Siberia, this change in climate would have caused the steppe tundra vegetation to change to marshy, lake-dotted tundra or to true forests, where mammoths were no longer able to find food.[42] In North America, mammoths also disappeared when the climate warmed. Between 15,000 and 10,000 years B.P., the patchwork of grasses and vegetation they lived on was replaced by swampy tundra and conifer forests, uninhabitable for them. Dale Guthrie, a specialist in paleobiology at the University of Fairbanks, has carefully studied the ecological changes in central Alaska between the Pleistocene and today.[43] These changes in vegetation would have been accompanied by climatic phenomena such as blizzards or mud slides (ice that had lost its solidity, torrents of mud)[44] that would have caused the death of large mammals.

The contrary hypothesis, which proposes the brutal extermination of Quaternary fauna by man, has found fervent disciples in the last three decades and is now the dominant, if not the "official," theory. This human action explanation gives renewed glory to Bernard Palissy's old theory that marine species disappeared because they were overfished by man. In 1926 an amateur English naturalist named George Bassett Digby[45] wove epic scenarios of mammoth hunts in Siberia modeled on the American buffalo hunts of historic times. But it was especially in the United States that the hypothesis was destined to have its greatest success. Since 1967 the prehistorian Paul Martin has been the most ardent defender of the "overkill" theory of a "massive extinction" of Pleistocene mammals by man.[46] According to Martin, this scenario works only in light of a very rapid extermination: the mammoths were destroyed in a "blitzkrieg," or lightning war. The especially short chronology of their extinction in America (between 11,000 and 10,800 years B.P.) seems to coincide with the arrival of man on the American continent. A *Homo sapiens* much like ourselves, a well-armed, skilled hunter with neither taboo nor restraint, entered a territory rich in game and committed a massacre.

But a more recent examination of the warring theories suggests that both scenarios have weaknesses and uncertainties. The long-dominant blitzkrieg theory is today at an impasse because it is difficult to find proof for it. When Donald Grayson in 1984 summarized two decades of controversies, he remarked that the theory has an unfortunate tendency to incorporate all objections to it, which leaves no room for refutation.[47] For example, Paul Martin writes, "if the extinctions were sudden, which seems to have been the case, it is not surprising that fossil evidence is insufficient to reveal many details."[48] Can such reasoning, which uses negative evidence to establish the truth of a theory, have real scientific consistency? Are the few archaeological traces of the Clovis culture which testify to mammoth hunting in North America, enough to prove the extermination of large numbers of individuals and a great number of species? They concern only the mammoth and the bison, and not all the other species that disappeared at the same time at the end of the Quaternary. And the bison is not extinct. . . .

In some parts of central Europe, where "civilizations of the mammoth"[49] developed—Czechoslovakia, Poland, in the Russian sites of the Don Valley and the Caucasus, and in Ukraine—large-scale hunting may have played a determining role in the extinction of a species that was already weakened and less numerous. Mammoths started to decline 14,000 years ago and disappeared from those regions around 12,000 years ago. But the mammoth bones in those sites probably accumulated year after year for centuries, so how can one speak of a "blitzkrieg"? Finally, can we hold man responsible for the disappearance of mammoths in the frozen wastes in the great Siberian or Alaskan north, where there is little evidence of killings, of hunting and butchering sites, and few archaeological signs of human activity?

The extinction of the mammoth fits with what we know of the arrival of the first humans on the American continent, but the chronology is different on other continents. Mammoths became extinct at different periods in central Europe (14,000 to 12,000 years B.P.), in China (14,000 years B.P.), and in Siberia (9,000 years B.P.)—and even at intervals of several millennia. How does the fact that man could have been the contemporary of these animals in Europe accord with the idea of a rapid extermination? Narratives must therefore vary according to geographic locations. Do any of these scenarios take the complexity of the phenomena into account? And is it really necessary to choose among them? "If you consider the immense distribution of the mammoth, which covers a wide diversity of natural conditions, it is impossible to affirm that a single natural factor caused its extinction," writes the Russian paleontologist Nikolai Vereshchagin.[50]

Some have tried to combine the different hypotheses. The initial aggression may have come from man, and the animals eventually became extinct when driven to inhospitable regions of the earth. Or, to the contrary, the animals were weakened by changes of climates, and man's intensive hunting dealt them a death blow. Brutal or progressive extinction? Catastrophism or gradualism? Human action or the effect of changes in the milieu? The scenarios have become more precise and more expansive, and some facts are better known, but questions remain, and they lead us back today to the very first inquiries raised by paleontology.

One recent discovery has rekindled the debate in spectacular fashion. In 1993 Russian scientists reported discovering mammoths that became extinct 3,700 years ago—barely 1,700 years b.c.—on Wrangel Island, 140 miles off the coast of eastern Siberia.[51] This extends by more than 6,000 years the survival of a species we thought had vanished from the surface of the earth by the end of the Pleistocene. Mammoths are symbols of Quaternary periods and witnesses to the adventures of the great Paleolithic hunters. That they could have survived to be the "contemporaries of the pharaohs of Egypt"[52] puts the issue of their extinction in a new light. The Wrangel Island mammoths are almost "living fossils," like the coelacanth discovered off Madagascar in 1938. This revives the debate on the causes for such longevity, and here again, various theories collide. "Extermination" supporters imagine that the mammoths survived on this island far from the coast because their predator, man, was not to be found there. No trace of human occupation has been discovered on Wrangel Island—another (negative) argument in favor of the destructive action of man. But the presence of mammoths at such a recent epoch could also support the climatic and environmental theory. The vegetation on Wrangel Island today represents a "relic" of the Pleistocene steppe tundra, which could explain the survival of mammoths into historical times in a favorable environment. Still, one has to wonder about the reasons for their extinction. Must we bring in human intervention—paleo-Eskimos from Siberia or Alaska who crossed the sea to hunt mammoths for their meat and ivory, thereby exterminating a small, weakened population that was limited geographically to one or several islands? Or maybe these mammoths, surviving in difficult conditions, became extinct here like "everywhere else"—the hypothesis put forth by the team of Russian scholars.

> Although oppressed and dwarfed, woolly mammoth could exist on Wrangel Island until at least 3,700 years ago. There is no evidence that it was hunted by man. The extraordinary fact is not that mammoth eventually became extinct on Wrangel like elsewhere, but that, together with some relics of tundra-steppe

flora, it was able to survive the early Holocene environmental revolution in this Arctic island refuge.[53]

Standing just six feet high at the shoulder and weighing "only" two tons (compared to the usual estimated average of ten to fourteen feet and six tons), the dwarf mammoths of Wrangel Island were fragile survivors. Their dwarfism can be explained by the environment and biogeography: shrinking is often linked to survival on islands, where conditions are difficult and food resources more limited.[54] This explanation can be verified by the fact that one finds the bones of normal-sized mammoths at the same site from a more ancient epoch, about 12,000 years before the present. The dwarfing would have begun at the time when the island was cut off from the continent because of the warming of the climate and a rise in sea level.

> Wrangel Island could still have been joined with the mainland as late as 13,000 yr BP, but by 12,000 yr BP that connection had been broken. The normal size of Wrangel mammoth teeth dating from the Pleistocene agrees with these estimates: it seems that the local population was not isolated from the main range until 12,000 yr BP. Later there is a gap in the mammoth record on the island until 7,000 yr BP, when the dwarfing had already occurred. This seems to have been enough time for insular dwarfing, comparable to the dwarfing of red deer on the island of Jersey (UK), which took less than 6,000 years.[55]

Finally, the dwarf mammoths of Wrangel Island raise a new and unexpected question: Could the traditions and the "mammoth myths" among the indigenous peoples of northern Siberia or Alaska have come from actual encounters with living animals? Do any of our contemporaries' images and stories reflect a *direct* memory of the life of these prehistoric animals? It is a troubling question because it blurs the traditional view—inherited from Cuvier—of a world of prehistory divided into epochs that are irreducibly different and distant: it would mean that the mammoth, once a symbol of a vanished past, would henceforth also belong to our own history.

The question of extinction can only be resolved by telling stories in which hypotheses and fictional constructions are combined with the few known "facts." Over the last centuries, we have had many accounts that seem to raise the same questions and the same images: climatic catastrophes, slow change

A paleontologist contemplating the "mammoth graveyard" at Berelyokh in eastern Siberia. (N. K. Vereshchagin photo.)

of the environment, action of man, constitutional weakness. But extinction scenarios, however logical or believable, cannot be proven. One cannot conduct experiments on the past, and the time periods involved are out of scale with the ones we experience. These scenarios are related in some way to the construction of a fiction—or of a tragedy—in the temporal sequence they propose, the various "actors" they bring on stage, and the forms of temporality, causality, and destiny they put into play. Constructing these "scenarios" is not far from what Descartes called "forging hypotheses"[56] in the exposition of his *Système du monde*.

The duel is between abstract "models" pushed to their extremes, logical fictions that are necessarily stories, because they take place over time, even if compressed to a "blitzkrieg." Anthropomorphism often dominates the way we approach the question, but how could it be otherwise? Explaining the disappearance of the mammoth means accounting for the inevitable death of a species, the extinction of a gigantic, powerful mammal that was a contemporary of man during the longest part of our history. Whether the mammoth disappeared because of its own biological tendencies, for ecological reasons, by the action of man, or by their combined effect, each hypothesis tends to make us meditate on our own fate. No doubt our obsessing over the question today is linked to real anxieties about our survival. What will be the end of humanity? Will it be linked to our "hyper-specialization" (human beings becoming pure brain, with atrophied members and body) or to an ecological catastrophe? Will it be the work of humankind itself? The question remains open, but the endlessly renewed narratives about the end of the mammoths, the extinction of dinosaurs or of Neanderthal man—our vanished brother from a still-recent past—though they are born of scientific motivation, are still a troubling reflection of our own questioning, of our own terrors.

CONCLUSION:
THE FUTURE OF PALEONTOLOGY

The history of science, and in particular the history of paleontology, has usually been conceived of as that of heroes and discoveries or of the development of ideas, problems, and methods.

By focusing on a particular object (mammoth bones and the fossil species to which they belong), this book has taken a new way to explore that history. In choosing a canonical object, one of paleontology's totem animals, as a reference point, it has tried to shed light on the variations in scientific outlook over the course of history, to measure the changes in methods and theories, and to follow the debates and controversies. This approach has also let us examine men and their discoveries, the ideas and circumstances that led to them, the practices of excavations and fieldwork, and the material, institutional, political—not to mention economic—conditions of research, and to describe the systems and stories that have been built on fossil remains for more than three centuries.

The history of paleontology is not entirely contained within the history of the mammoth, of course. Specific questions are raised by the study of fossil plants and invertebrates, the great Mesozoic reptiles, the first mammals, and by the study of prehistoric humans and their cultures. But the path this book has taken gives a fairly clear picture of the development of a field of knowledge that has undergone profound changes since the end of the Renaissance. At the close of the seventeenth century, the question of the origin of fossils began to split off from theological issues, and thus opened for many naturalists the dimension of an immense period of time and a history of the earth prior to human existence. At the beginning of the nineteenth century, fossils were recognized as belonging to "lost species" and vertebrate paleontology became an institutionalized scientific discipline, with its own concepts and methods. But intense debates in France and England raised the question of the continuity and discontinuity of the history of life and of the earth. Between 1830 and 1860, the question of the evolution of living beings came to the fore, and discoveries on human origins began to attract attention. The year 1859 was a pivotal one, in which the Darwinian perspective of evolution was defined and the notion of fossil man accepted. In the United States during the first decades of the twentieth century, the synthetic theory of evolution linked the study of fossils with the new science of genetics and later with molecular biology. In the 1960s the rise of a mobilist geography and the acceptance of the theory of plate tectonics opened new fields to paleontological research and knowledge.

At each step of this history, the representations of the living world and of the earth were deeply transformed. The mutations of knowledge are marked by changes of theoretical paradigms, practices, and scientific institutions. They are linked to cultural, economic, and political transformations of the societies that produced them, but they also transformed these human groups' view of the world.

The history of paleontology does not only consist of breaks and discontinuities, however. Over the centuries, knowledge about the history of life and the earth has been immensely enriched by the increase in excavations and empirical discoveries; by the spread of fossil collections to natural history cabinets, museums, and galleries of evolution;[1] by the ever-growing number of reconstructed extinct species and genera; and by many publications, specialized periodicals, monographs, and works of synthesis. The progression has also been marked by a deepening of the questions asked. The capacity to reconstruct "lost species" was established long before a transformist theory put the succession of fossil fauna in evolutionary perspective, and evolution itself was conceived of and described several decades before its biological mechanisms were elucidated. In this way, the history of paleontology is marked by successive steps that extend and surpass previous knowledge.

Understanding the history of paleontology requires that we consider not only discoveries, practices, and theories, but also imaginary constructs, because the role of fiction is constant both in the development of this science and its relationship to the public. The play of the imagination accompanies the discoveries, interpretations, and reconstructions of extinct beings. It enters into the elaboration of theories in forging hypotheses and filling the gaps of stories and is essential to the spread of scientific knowledge and the needs of the general public.

Is there a reservoir of common themes and images that spring up in the discourse of science—or what passes for it—and also in the more diffused and popularized forms of knowledge? In the history of paleontology, ancient beliefs have endured or have recurred, such as belief in the gigantic size of past beings, the representation of a history of the earth and of life marked by catastrophes, the model of a linear evolution that culminates with the appearance of man, or that of cyclical time causing a recurrence of past events.

These images, whether they are conceived of as part of our inherent mental "schemas" or as the product of cultural traditions, are not inert and static, but dynamic. They lead to a never-ending quest: expeditions to the far corners of the world, a search for origins, or speculations about "missing links."

Imagination has a history. It is not only contemplative; it puts itself to the test of seeking and is transformed and even reshaped with new data. It is possible that the object of prehistory, a creature from past times, will open a new imaginary world and in some way create a new myth. The mammoth's popularity may be partly because it shares traits both of the elephant and the teddy bear, and partly because it embodies archaic schemes of imagination and sensitivity, images of fragility and power. But does it not also come from the fact that it is set in a particular universe, composed of migrations to the ends of the earth, frozen wastes and blizzards, relations with prehistoric men, epic or tragic scenarios of hunting and extinction? In the same way, the dinosaur fits into the mythical family of monsters, dragons, and sea serpents—though it would cheapen our image of the dinosaur to say that it only belongs to that mythical configuration. Dinosaurs have their own mythic constellation, their own saga and their imaginary personality, which makes them distinctive.

The sciences of prehistory occupy a special place in contemporary imagination precisely because they suggest or evoke such images. In an era increasingly dominated by the image, these sciences produce images and reconstitutions of very ancient times. The disciplines that study the prehistoric past of the living world are very broadly popularized, and the forms of popularization—novels, movies, comic strips, advertising—play a vital role in keeping them alive to the public. In all societies, myth speaks of origin. And paleontological knowledge has served, and perhaps still serves, as our origin myths. Is not the function of myths linked precisely to our belief in them? In many ways, prehistoric sciences give reality to dreams; they take the place of myths—they *create* myths. And, like myths, they raise the question of origin and change, questions that are constantly renewed and serve, like myths, to ensure social cohesion.

Paleontologists are explorers in a double sense, since they travel both through space and time. In his day Cuvier made use of this image for the prestige of the science he claimed to have founded. Today the paleontologist has become the hero of novels and movies and may even be the last true adventurer. But at a time when mass media play a pivotal role in the diffusion of knowledge, there is a risk of seeing scientists enter into legend to the point of becoming only mythic figures, and perhaps of being threatened with inexistence or extinction.

Paleontology, which was triumphant in the first half of the nineteenth century, in fact occupied a rather minor place in Darwin's works. Today new directions in research inspired by geophysics and molecular biology are over-taking the methods of traditional natural history. At a time when fossils may be less necessary as proofs of evolution, will the traditional descriptive and classifying approach to extinct beings soon be replaced by laboratory study of genetic distances between species? Does paleontology still have a place among contemporary sciences? Or are fossils today no more than images of dreams, the materialization of an abolished past?

As a discipline, paleontology—its institutions, the number and training of its scholars, its possibilities for field and laboratory research—is threatened nearly everywhere in the world. Yet it alone allows us to approach the evolution of the living world in its diversity, to know the aspect and history of each of the beings that have existed in their irreducible singularity. Fossils provide essential information in reconstituting the history of the earth. They raise the questions of the biogeographical history of species, the causes of extinctions, the trends of evolution, and the transformations of the living world. Through their study, questions may arise about the meaning of the destiny of life and of humanity on the geological scale.

For all these reasons, as a scientific discipline paleontology itself not only has a past; it also has a future. What new stories will be told tomorrow about the mammoth? The journey we have followed here is far from being finished. . . .

We must now close the book, and let the mammoth make its way towards its destiny.

The Berezovka mammoth arrives, on sleds drawn by ponies, at one of the stages in its journey. (V. E. Garutt personal collection.)

Preface to the American Edition

1. On the history of paleontology, see Robert Chambers, *Vestiges of the Natural History of Creation and Other Evolutionary Writings*, ed. James A. Secord (Chicago: Chicago University Press, 1994); Martin J. S. Rudwick, *Georges Cuvier, Fossil Bones, and Geological Catastrophes* (Chicago: University of Chicago Press, 1997); Adrienne Mayor, *The First Fossil Hunters: Paleontology in Greek and Roman Times* (Princeton: Princeton University Press, 2000); Tim Murray, ed., *Encyclopedia of Archaeology*, Vol. 1, *The Great Archaeologists* (Santa Barbara, Calif.: ABC-Clio, 1999); C. Cohen, C. Blanckaert, J. L. Fisher, and P. Corsi, eds., *Le Muséum au premier siècle de son histoire* (Paris: Editions du Muséum National d'Histoire Naturelle, 1997). The history and even the "biography" of objects has become a new genre in the history of science. See C. Cohen, "De l'histoire de l'objectivité scientifique à l'histoire des objets de science," in *Des sciences et des techniques: Un Débat, Cahiers des Annales 45*, ed. R. Guesnerie and F. Hartog (Paris: Editions de l'EHESS, 1998), 149–56; and Lorraine Daston, ed., *Biographies of Scientific Objects* (Chicago: University of Chicago Press, 1999). For biographical monographs, see Adrian Desmond, *Huxley: From Devil's Disciple to Evolution's High Priest* (Cambridge, Mass.: Perseus Books, 1994) (as well as several biographies of Darwin); Nicolaas Rupke, *Richard Owen: Victorian Naturalist* (New Haven: Yale University Press, 1994); Goulven Laurent, ed., *Lamarck* (Paris: Editions du CTHS, 1997); Derek J. Blundell and A. C. Scott, eds., *Lyell: The Past Is the Key to the Present* (London: Geological Society Special Publications, 1998). For the history of geology, see David R. Oldroyd, *Thinking about the Earth: A History of Ideas in Geology* (London: Athlone, 1996); Nicoletta Morello, ed., *Volcanoes and History: Proceedings of the 20th INHIGEO Symposium* (Genoa: Brigati-Genova, 1998).

2. See M. Noro, R. Masuda, I. A. Dubrovo, M. C. Yoshida, and M. Kato, "Phylogenetic Inference of the Woolly Mammoth, *Mammuthus primigenius*, Based on Complete Sequences of Mitochondrial Cytochrome b and 12S Ribosomal RNA Genes," *Journal of Molecular Evolution* 46 (1998): 314–24. See also T. Osawa, S. Hashaui, and V. M. Mikhelson, "Phylogenetic Position of Mammoth and Steller's Sea Cow within *Tethytheria* Demonstrated by Mitochondrial DNA Sequences," *Journal of Molecular Evolution* 44 (1997): 406–13. Régis Debruyne, *Phylogénie moléculaire des Elephants (Mammalia, Proboscidea) et position du Mammouth laineux*, Diplome d'Etudes Approfondies, "Biodiversité: Génétique, Histoire et Mécanismes de l'Evolution," Universités Paris VI, Paris VII, Paris XI, INA-PG, MNHN, September 2000. See also V. Barriel, R. Debruyne, and P. Tassy in *Journal of Molecular Evolution* (forthcoming).

3. Emmauel Gherbrant, Jean Sudre, and Henri Capetta, "A Palaeocene Proboscidean from Morocco," *Nature* 383 (September 1996): 69–70.

4. See Francis Latreille and Bernard Buigues, *Mammouth* (Paris: Robert Laffont, 2000).

5. See chapter 10, page 203.

6. Mayor, *The First Fossil Hunters*, 249.

Preface

1. A. Leroi-Gourhan, "Le Mammouth dans la zoologie mythiques des Eskimos," *Le Fil du temps* (Paris: Fayard, 1983), 37. The article was first published in *La Terre et Vie* 5, no. 1 (1935).

Introduction

1. Among the classics of paleontology history, see Karl von Zittel, "Geschichte der Geologie und Paläeontologie bis Ende des 19 Jahrhunderts," *Geschichte der Wissenschaft in Deutschland*, vol. 23 (Munich: Oldenbourg, 1899); John Greene, *The Death of Adam: Evolution and Its Impact on Western Thought* (Ames: Iowa State University Press, 1959); Martin J. S. Rudwick, *The Meaning of Fossils: Episodes in the History of Paleontology*, 2nd ed. (Chicago: University of Chicago Press, 1976); Paolo Rossi, *The Dark Abyss of Time*, trans. Lydia G. Cochrane (Chicago: University of Chicago Press, 1984); Eric Buffetaut, *Des fossiles et des hommes* (Paris: Laffont, 1991), trans. of *A Short History of Vertebrate Palaeontology* (London: Croom Helm, 1987); Helmut Hölder, *Kurze Geschichte der Geologie und Paläeontologie* (Berlin: Springer-Verlag, 1989).

2. See in particular Adrian Desmond, *Archetypes and Ancestors: Paleontology in Victorian London 1850–1875* (Chicago: University of Chicago Press, 1982); and *The Politics of Evolution: Morphology, Medicine, and Reform in Radical London* (Chicago: University of Chicago Press, 1989); Stephen Jay Gould, *Time's Arrow and Time's Cycle* (Cambridge: Harvard University Press, 1987); Martin J. S. Rudwick, *The Great Devonian Controversy* (Chicago: University of Chicago Press, 1985) and *Scenes from Deep Time* (Chicago: University of Chicago Press, 1992); Nicolaas A. Rupke, *The Great Chain of History: William Buckland and the English School of Geology (1814–1849)* (New York: Oxford University Press, 1983); and *Richard Owen: Victorian Naturalist* (New Haven: Yale University Press, 1994).

3. See, for example, Stephen Jay Gould, *Wonderful Life* (New York: Norton, 1989).

4. See the titles of works by Stephen Jay Gould, such as *The Panda's Thumb* (New York: Norton, 1980) and *The Flamingo's Smile* (New York: Norton, 1985).

5. Niles Eldredge, "Cladism and Common Sense," in *Phylogenetic Analysis and Paleontology*, ed. J. Cracraft and N. Eldredge (New York: Columbia University Press, 1979), 192–95.

Chapter 1

1. Mikhail Ivanovich Adams, "Some Account of a Journey to the Frozen Sea and the Discovery of the Remains of a Mammoth," *Philos. Magazin* 29 (Tilloch, Oct.–Dec. 1807 and Jan. 1808): 141–43.

2. See Rudwick, *Scenes from Deep Time*, 158, 165.

3. On the images and reconstitutions of mammoths, see N. K. Vereshchagin and A. N. Tikhonov, *Eksterier Mamonta* (U.S.S.R. Academy of Sciences, Siberia Division, August 1990).

4. See Eugen W. Pfizenmayer, *Siberian Man and Mammoth*, trans. Muriel Simpson (London: Blacking, 1939).

5. On mammoths in Paleolithic art, see Capitan, Breuil, and Peyrony, "La Caverne de Font-de-Gaume aux Eyzies (Dordogne)" (Monaco, 1910); M. Sarradet, "Font-de-Gaume en Périgord" (Périgueux, 1977); C. Maska, H. Obermaier, H. Breuil, "La Statuette de mammouth de Předmost," *L'Anthropologie* 23 (1912): 3–4; L. R. Nougier and R. Robert, *Rouffignac, la grotte aux cents mammouths* (Paris, 1957); L. R. Nougier and R. Robert, *Rouffignac, ou La Guerre des mammouths* (Paris, 1957); G. Bosinski, "The Mammoth Engravings of the Magdalenian Site of Gönnersdorf (Rhineland, Germany)," in *IIIe Colloque de la Société Suisse des sciences humaines* (Geneva, 1979) ; B. and G. Delluc, "Les Grottes ornées de Domme, Dordogne: La Martine, le mammouth et le prisonnier," *Gallia Préhistoire* 26, no. 1 (1983). See also Henri Breuil, *Quatre cents siècles d'art pariétal* (Mame: 1952; reprint, 1972); André Leroi-Gourhan, *La Préhistoire de l'art occidental* (Paris: Mazenod, 1965); Zoia Abramova, *Paleolithic Art in the U.S.S.R.*, trans. Catherine Page, ed. Chester Chard (Madison: University of Wisconsin Press, 1967).

6. V. P. Lyubin, "The Representation of the Mammoth in Paleolithic Art" (in Russian), *Sovriemiennaia archeologia* (St. Petersburg, 1991), 1:20–42.

7. Zoia Abramova, "Les Correlations entre l'art et la faune dans le Paléolithique de la plaine russe (la femme et le mammouth)" (Correlations between art and fauna on the Paleolithic Russian plain [woman and mammoth]) (1979), 334.

8. See "Mamontovaia fauna rousskoi ravniny i vostochnoï siberi" (Mammoth fauna of the Russian plains and northern Siberia), *Proceedings of the St. Petersburg Zoological Institute* (Leningrad, 1977), 70.

9. See the discussion by Rudwick in *Scenes from Deep Time*.

10. See Louis Figuier, *La Terre avant le Déluge*, 6th ed. (1867; reprint, Paris, 1961); trans. as *The Earth before the Deluge*, ed. Henry Bristow (London: Chapman & Hall, 1867).

11. See Emile Michel, "Les Peintures décoratives de M. Cormon au Muséum," *Revue d'art ancien et moderne* 3 (1898).

12. For a discussion of Knight's life and work, see Sylvia Massey Czerkas and Donald F. Glut, *Dinosaurs, Mammoths and Cavemen: The Art of Charles R. Knight* (New York: Dutton, 1982).

13. The painting is in the American Museum of Natural History in New York. Another one on the same subject was painted in 1939.

14. On Burian, see Vladimir Prokop, *Zdenek Burian a paleontologie* (Prague: Vydal Ustredni Ustav Geologicky, 1990). See also the exhibit catalog for *Peintres d'un monde disparu. La Préhistoire vue par des artistes de la fin du XIXe siècle à nos jours* (Musée départemental de préhistoire de Solutré, June 22–October 1, 1990).

15. Josef Augusta and Zdenek Burian, *A Book of Mammoths*, trans. Margaret Schierl (London: Hamlyn, 1963).

16. See Louis Figuier, *La Terre avant le Déluge* and *L'Homme primitif* (Paris, 1870; reprint, 1870, 1873, 1882); Camille Flammarion, *Le Monde avant la création de l'homme. Origines de la terre, origines de la vie, origines de l'humanité* (Paris, 1886).

17. Jules Verne, *Voyage au centre de la terre* (Paris, 1864).

18. Edmond Haraucourt, *Daah, le premier homme* (Paris: 1914), reprinted with a preface by Geneviève Guichard (Paris: Arléa, 1988).

19. Adrien Cranile [Adrien Arcelin], *Solutré, ou Les Chasseurs de rennes de la France centrale. Histoire préhistorique* (Paris, 1872).

20. See J. H. Rosny Aîné, *Romans préhistoriques* (Paris: Laffont, 1985).

21. Jean Auel, *The Mammoth Hunters* (New York: Bantam, 1986), 19.

22. Jack London, "A Relic of the Pliocene," *The Faith of Men and Other Stories* (New York: Macmillan, 1904).

23. Max Bégouën, *Quand le mammouth ressuscita* (Paris: Hachette, 1928), 104. I thank Michel Alain Garcia for sharing this wonderful book with me.

24. See Michael Crichton, *Jurassic Park* (New York: Knopf, 1990), and Stephen Spielberg's 1993 movie *Jurassic Park*.

25. See Pierre Gouletquer, "La Préhistoire de bande dessinée: Mythes et limites," in *Historiens-Géographes* 318:371–83.

26. See Aidans, *Tounga, le maître des mammouths* (Brussels: Lombard, 1982).

27. R. Lecureux, A. Cheret, and C. Cheret, *Les Nouvelles aventures de Rahan, fils des âges farouches;* and *Rahan contre le temps* (Brussels: Novedi, 1991).

Chapter 2

1. *De Civitate Dei* (The city of God), book 15, *Basic Writings of Saint Augustine*, ed. Whitney J. Oates (New York: Random House, 1948) 2:286.

2. The notion of time here is no longer seen in terms of the cyclical model of an eternal return. The Christian representation of time carries a new dimension, that of history and eschatology. On Christianity and history, see Lucien Febvre, "Vers une autre histoire," *Revue de métaphysique et de morale*, nos. 3–4 (1949): 225–48.

3. Augustine, *De Civitate Dei*, 284.

4. On this topic, I have borrowed a great deal from Jean Céard, "La Querelle des géants et la jeunesse du monde," *Medieval and Renaissance Studies* 8, no. 1 (spring 1978): 37–77.

5. See Lorraine Daston, "Marvellous Facts and Miraculous Evidence in Early Modern Europe," *Critical Inquiry* 18 (fall 1991): 93–124.

6. Pliny the Elder, *Natural History*, 7:16.

7. See Rudwick, *The Meaning of Fossils*.

8. Augustine, *De Civitate Dei*, 284.

9. Ibid.

10. See Othenio Abel, *Vorzeitliche Tierreste im Deutschen Mythus, Brauchtum und Volksglauben* (Jena: Gustav Fischer, 1939); Buffetaut, *Des fossiles et des hommes* and Antoine Schnapper, *Le Géant, la licorne, la tulipe* (Paris: Flammarion, 1988). On this question, see also Antoine Schnapper, "Persistance des géants," *Annales ESC*, no. 1 (Jan.–Feb. 1986): 177–200.

11. See Raymond Vaufrey, *Les Eléphants nains des îles méditerrannéennes* (Paris: Masson, 1929); and Abel, *Vorzeitliche Tierreste*.

12. Céard, "La Querelle."

13. Pliny, *Natural History*, 7:16.

14. Boccaccio, *De genealogica deorum gentilium*, 4:68; trans. as *Genealogies of the Pagan Gods*.

15. Giuseppe Olmi, *L'inventario del mondo. Catalogazione della natura e luoghi del sapere nella prima età moderna* (Bologna: Società editrice il Mulino, 1992); my translation.

16. Ibid., 165–66.

17. Figuier, *La Terre avant le Déluge*, 336.

18. Céard, "La Querelle," 47.

19. Quoted in Schnapper, *Le Géant*, 98.

20. Nicholas Habicot, *Antigigantologie, ou Contrediscours de la grandeur des géants* (Paris, 1618), 58–59.

21. See Rossi, *The Dark Abyss of Time.*

22. Ulisse Aldrovandi, *Museum metallicum, in libros IV distributum* (Bologna, 1648), a posthumous work edited by Bartholomeo Ambrosinus.

23. Athanasius Kircher, *Mundus subterraneus* (Rome, 1665), vol. 2.

24. *The Notebooks of Leonardo da Vinci*, transl. and intro. by Edward MacCurdy (London, 1938).

25. Bernard Palissy, *Discours admirable des eaux et des fontaines* (1580), in *Oeuvres* (1880), 447. On Palissy, see L. Audiat, *Bernard Palissy. Etude sur sa vie et ses travaux* (Paris, 1868).

26. Palissy, *Discours*, 334.

27. "[. . .] j'ai jugé à la fois opportun et nécessaire de rassembler quelques exemples très antiques qui font foi de la vérité et en meme temps de raconter les choses que j'ai vues de mes propres yeux, en les réunissant à l'autorité de ces écrivains très antiques et au jugement de l'écriture sainte." (I judged it both opportune and necessary to gather a few examples that attest to the truth, and at the same time to describe things I have seen with my own eyes, adding them to the authority of these very ancient writers and the judgment of Holy Writ.) T. Fazello, *De rebus siculis decades duae . . .* (Palermo, 1558), translated into Italian (Venice, 1573), 33. Quoted in Schnapper, *Le Géant*, 182.

28. Goropius [Jan van Gorp], *Gigantomachie* (1559).

29. Jean Riolan, *Gigantologie, discours sur la grandeur des géants, où il est démontré, que de toute ancienneté les plus grands hommes & géants, n'ont été plus hauts que ceux de ce temps* (Paris: Adrien Perier, 1618), 88.

30. Ibid., 87.

31. Jules Michelet, *Histoire de France: Le Moyen âge* (1833; reprint, Paris: Laffont, 1981), 44–47.

32. Paul Orose, *Ystoire des Romains*, ed. V. Vérard (Paris, 1509). First translation into French, 1491.

33. Nicholas Habicot, *Gigantéostologie, ou Discours des os d'un géant* (Paris, 1613), 57.

34. Nicholas Habicot, "a celebrated anatomist, was born around 1550 at Bonny, in the Gâtinais," according to Michaud's *Biographie universelle* 18 (Paris, 1857), 308. He was a surgeon at the Hôtel-Dieu de Paris, served with the military, and was on the faculty at the college de Saint-Côme.

35. Riolan also deserves credit for suggesting the creation of the Jardin botanique du Roi, which later became the Jardin des Plantes in Paris.

36. Habicot, *Gigantostéologie*, 22–23.

37. Ibid., 33.

38. Ibid., 17.

39. Nicholas Habicot, *Réponse à un discours apologétic touchant la vérité des géants* (Paris, 1615).

40. Riolan, *Gigantologie.*

41. Ibid., 99.

42. Jean Riolan, *Gigantomachie, pour répondre à la gigantostéologie* (Paris, 1613), 9.

43. Riolan, *Gigantologie*, 64.

44. Ibid., 34–35.

45. Ibid., 35.

46. Ibid., 41.

47. Ibid., 43.

48. "For a man who had never seen an elephant skeleton, Riolan showed considerable skill in demonstrating that the bones must have come from that animal." Georges Cuvier, *Recherches sur les ossemens fossiles de quadrupèdes*, 3rd ed. (Paris, 1825), 1:102.

49. Georges Cuvier, *Recherches sur les ossemens fossiles de quadrupèdes*, 1st ed. (Paris, 1812), 1:90. On Hannibal's crossing the Alps with elephants, see Francis de Conninck, *La Traversée des Alpes par Hannibal selon les écrits de Polybe* (Montélimar: Ediculture, 1992).

50. Riolan, *Gigantologie*, 44.

51. Ibid., 47–48.

52. Ibid., 32.

53. "Riolan's weakness is to hesitate between his explanations, from one pamphlet to the next, and almost from one page to the next," writes Schnapper in "Persistance," 101.

54. See A. Koyré, "L'Hypothèse et l'expérience chez Newton," in *Etudes newtoniennes* (Paris, 1967), 53–84.

55. Léonard Ginsburg, "Nouvelles lumières sur les ossements fossiles autrefois attribués au géant Theutobochus," *Annales de paléontologie* 70 (Paris, 1984): 181–219. Quoted in Buffetaut, *Des fossiles et des hommes*.

56. "On accusa ce chirurgien d'avoir fait faire sa brochure par un jésuite de Tournon, qui avoit forgé l'histoire du sepulchre et de l'inscription; ses prétendues medailles portaient des lettres gothiques et n'avoient rien de romain. Il ne paroit pas qu'il se soit justifié de cette imposture." (This surgeon was accused of having his pamphlet written by a Jesuit from Tournon who made up the history of the sepulchre and the inscription; his so-called medals bore Gothic letters, and there was nothing Roman about them. He does not seem to have justified this imposture.) In vol. 1 of his *Recherches* (1812), Cuvier quotes book 3 of Gassandi's *Vie de Peiresc*, and in vol. 5 of his *Oeuvres*, 280.

57. Céard, "La Querelle," 76.

58. "On trouve partout dans des grottes ou des fosses souterraines, des dents, des tibias, des côtes d'une grandeur immense, dont beaucoup disent que ce sont des os de géants. . . . On trouve des os de plusieurs genres dans les creux souterrains, comme dans les vieilles fosses de tombeaux. Parmi lesquelles on dit qu'il y a des tibias d'une grandeur énorme, et des ossements semblables à ceux des hommes qui auraient appartenu à des géants." (Throughout caves or underground pits one finds teeth, tibias, and ribs of immense size, which many say are the bones of giants. . . . One finds several kinds of bones in subterranean hollows, such as old tombs. Among them, it is said, are enormous tibias and bones resembling those of man, which might have belonged to giants.) Kircher, *Mundus subterraneus*, book 8, sec. 2, chap 4, "De ossium et cornuum subterraneorum genesi" (Of the subterranean generation of bones and horns), 53.

59. G. W. Leibniz, Hanover ms. LH 37, 4, Ff. 14–15 (in French; undated). I thank Michel Fichant for giving me a copy of this unpublished and hitherto unknown manuscript.

60. See Schnapper, "Persistance."

61. See Schnapper, *Le Géant*.

62. See Maupertuis, *La Vénus physique* (Paris, 1749); trans. by Simone Boas as *The Earthly Venus* (New York: Johnson Reprint, 1966).

63. See Henry Fairfield Osborn, *The Origin and Evolution of Life* (London, 1918), and *Titanotheres of Ancient Nebraska* (Washington, 1929).

64. Franz Weidenreich, *Apes, Giants, and Man* (Chicago: University of Chicago Press, 1946).

Chapter 3

1. *Cyranides*, attributed to Hermes Trismigestus, a gnostic text written between A.D. 227 and 400. See F. de Mély, *Les Lapidaires de l'antiquité et du Moyen âge* (Paris, 1898–1902).

2. See Odell Shepard, *The Lore of the Unicorn* (London: Shepard, Allen & Unwin, 1930); Roger Caillois, *Le Mythe de la licorne* (Paris: Fata Morgana, 1991).

3. See, for example, Pierre Belon, *Observations de plusieurs singularitez et choses memorables de divers pays estranges* (Paris, 1553) 1:14; *Discorso de Andrea Marini, medico, contra la falsa opinione dell'Alicorno* (Venice, 1566).

4. See Schnapper, *Le Géant, la licorne, la tulipe*.

5. See, for example, *L'Alicorno, discorso dell'excellente medico et filosofo M. Andrea Bacci, nel quale si tratta della natura dell'alicorno & delle sue virtu excellentissime* (Florence, 1573; first ed., in Latin, 1566).

6. See Shepard, *Lore of the Unicorn*, chap. 4, "The Battle of Books."

7. See especially Conrad Gesner, *Historia animalium* (Frankfurt, 1551); Ulisse Aldrovandi, *De quadrupedibus solipedibus* (Bologna, 1639); Jerome Cardan, *De subtilitate libri XXI* (Nuremberg, 1550); Laurens Catelan, *Histoire de la nature, chasse, vertus, propriétéz et usage de la lycorne* (Montpellier, 1624).

8. *Discours d'Ambroise Paré, conseilleur et premier chirurgien du Roy, à scavoir: de la mummie, de la licorne, des venins et de la peste* (Paris, 1582).

9. See Thomas Bartholin, *De unicornu observationes novae* (Poitiers, 1645).

10. Athanasius Kircher, *Mundus subterraneus* (1665), 2:63; my translation.

11. G. W. Leibniz, *Protogaea sive de prima facie Telluris et antiquissimae Historiae vestigiis in ipsis naturae Monumentis Dissertatio ex schedis manuscriptis Viri illustris in lucem edita a Christiano Ludvico Scheidio* (Göttingen, 1749).

Translator's note: Thanks to Roger Ariew of Virginia State University for his valuable help in translating the *Protogaea* text.

12. Ibid.

13. Otto von Guericke, *Experimenta nova magdeburgica de vacuo spatio* (Amsterdam, 1672), 5:155.

14. Here is von Guericke's text: "Accidit quoque ipso anno 1663 Quedlinburgi, quod in monte, quem vulgo den Zeuniken-berg vocant, ubi materia calcis effonditur, & quidem in ejus quadam rupe, repertum est Sceleton Unicornis, in postiore corporis parte, ut bruta solen, reclinatum, capite vero sursum elevato, ante frontem gerens longe extensum cornu crassite cruris humani, atque ita secundum proportionem longitudine quinque fere elnarum. Animalis quidem hujus sceleton primum ex ignorantia fuit contritum, & particulatim extratum, donec caput una cum cornu & aliquibus costis, spina dorsi, atque ossibus, Reverentissimae Principi Abbatissae, ibidem degenti, fuerit traditum. Ex quibus jam dictis, simul concludi potest, tellurem, longo temporis tractu, incrementum adsumere, specimenque vegetabilis augmenti prae se ferre." Ibid.

15. Leibniz, *Protogaea*, chap. 36.

16. Othenio Abel, *Geschichte und Methode des Rekonstruktion vorzeitlicher Wirbeltiere* (Jena: Gustav Fischer, 1925).

17. Leibniz includes a summary of *Protogaea* in *Theodicae* 3:242–45.

18. Yvon Belaval, *Leibnitz, initiation à sa philosophie* (Paris: Vrin, 1975), 157.

19. See L. Davillé, *Leibnitz historien* (Paris, 1909).

20. Cited in Gunther Scheel, "Leibnitz historien," in *Leibnitz* (Paris: Aubier-Montaigne, 1968), 55–56.

21. Leibniz, *Protogaea*, chap. 48.

22. Ibid.

23. See Jon Elster, *Leibnitz et la formation de l'esprit capitaliste* (Paris: Aubier-Montaigne, 1957), chap. 3, "Les Mines de Hartz," 85.

24. As early as 1683, Leibniz declared having discovered in rock formations, minerals, and fossil imprints "things that are so far from common opinion and yet so easy to explain by purely mechanical reasons that I attribute the lack of authors who have written about them only to the superficial way such matters have been treated and to certain prejudices held by miners that writers have adopted without discussion." In this inquiry, which was based on Leibniz's experience in the Harz mines, one can see the thinking that would later be elaborated in *Protogaea*.

25. Agricola [Georg Bauer], *De re metallica libri VII* (Basel, 1556), and *De natura fossilum* (Basel, 1546).

26. See Giovanni Solinas, "La *Protogaea* di Leibnitz ai margini della rivoluzione scientifica," in *Saggi sull'illuminismo* (Cagliari: Instituto di filosofia, 1973), 7–70.

27. Agostino Scilla, *La vana speculazione disingannata dal senso, Lettera risponsiva . . . circa i corpi marini che petrificati si trovano in varii luoghi terrestri* (Naples, 1670).

28. See Gabriel Gohau, *Une Histoire de la géologie* (Paris: Seuil, 1990), 67–68; trans. by Albert Carozzi and Marguerite Carozzi as *A History of Geology* (New Brunswick, N.J.: Rutgers University Press, 1990).

29. See Leibniz's March 22, 1714, letter to Louis Bourguet, in C. I. Gerhardt, *Die philosophischen Schriften von Gottfried von Leibniz* (Berlin, 1886), 3:565–66.

30. Rossi, *The Dark Abyss of Time*, 55.

31. "A letter in the Hanover archives dated January 1678, which is obviously inspired by Steno's *Prodromus* is entitled *Facies terrae*," writes Jacques Roger in "Leibnitz et la théorie de la Terre," *Leibnitz, 1646–1716. Aspects de l'homme et de l'oeuvre* (Paris: Aubier-Montaigne, 1968), 137–44.

32. Nicolaus Steno [Niels Steensen], *Canis Carcharidae Dissectum Caput et dissectus piscis ex Canum genere* (Florence, 1667); *De solido intra solidum naturaliter contento dissertationis Prodromus* (Florence, 1669).

33. Steno, *Canis Carcharidae*.

34. Leibniz, *Protogaea*, chap. 2.

35. G. W. Leibniz, Hanover ms. LH 37, 4, Ff. 14–15.

36. Leibniz, *Protogaea*, chap. 29.

37. Ibid., chap. 28.

38. Ibid., chap. 18.

39. Ibid, chap. 28.

40. *Histoire de l'Académie* (1706), 9–11.

41. Leibniz, *Protogaea*, chap. 18.

42. Ibid.

43. Etymology is the archaeology of language that, by studying word roots and their evolution into present-day words, allows one to discover the remote past of languages and of the people who spoke them. Leibniz compiled a German etymological glossary called *Archaeologus*.

44. Leibniz, *Protogaea*, chap. 10.

45. Ibid., chap. 9.

46. Ibid., chap. 18.

47. *Telliamed* was written between 1692 and 1720, after which it circulated in manuscript. The first edition was published in Amsterdam in 1748. See *Telliamed: or, Conversations between an Indian Philosopher and a French Missionary on the Diminution of the Sea*, trans. and ed. Albert Carozzi (Urbana: University of Illinois Press, 1968).

48. Leibniz, *Protogaea*, chap. 34.

49. Leibniz notes the existence of mammoth bones in Siberia, per Witsen, in *Protogaea*, chap. 34, "Of Bones, Maxillary Bones, Skulls, and Teeth of Various Sizes Found in Our Baumann Cave and Elsewhere."

50. This discovery was presented in a letter addressed by Wilhelm Tentzelius, historiographer to the duke of Saxony, to Antonio Magliabechi, counselor to the grand duke of Tuscany, *De sceleto elephantino a celeberrimo Wilhelmo Tentzelio Historigraphio ducali saxonica, ubi quoque Testaceorum petrificationes defenduntur . . .* (Urbini: Litteris Leonardi, 1697).

51. Letter to Thomas Burnett of Kemney, Hanover, July 17–27, 1696, in Gerhardt, *Die philosophische Schriften*, 3:184. The hypothesis had been advanced that the skeleton could be that of a "fossil unicorn," but Leibniz does not mention it. See *Des Unicornu fossilis oder gegrabenen Einhorn Welches in der Herrshafft Tonna gefunden worden Berfertiget von dem Collegio Medico in Gotha den 14 febr. 1696.*

52. Leibniz, *Protogaea*, chap. 34.

53. Ibid.

Chapter 4

1. See S. V. Ivanov, "The Mammoth in the Art of the Peoples of Siberian" (in Russian), in vol. 11 of the series by the museum of anthropology and ethnography (1949), 133–61. See also A. Leroi-Gourhan, "Le Mammouth dans la zoologie mythiques des Eskimos," *Le Fil du temps* (Paris: Fayard, 1983), 37.

2. Nicolaas Witsen, *Noord- en Oost Tartarie . . .* (Amsterdam, 1692–1705; reprint, 1785), 2:742–46.

3. P. Pekarski, *Naouk i littératoura v Rossii pri Pietre Viélikom* (Science and literature under Peter the Great) (St. Petersburg, 1862), 1:350–62; my translation.

4. Johann Georg Gmelin, *Reise durch Sibirien, von dem Jahr 1733 bis 1754* (Göttingen: 1751–52), trans. by Louis de Keralio as *Voyage en Sibérie*, 2 vols. (Paris, 1767). On Gmelin, see L. P. Velikovetz, *Johann Georg Gmelin* (Moscow: Nauka, 1990).

5. Henry H. Howorth, *The Mammoth and the Flood* (London, 1887), 78–80.

6. Eduard Vasilevich Toll, "A Geological Description of the Islands of New Siberia and the Principal Research Problems in Polar Regions," *Papers of the St. Petersburg Imperial Academy of Sciences*; my translation.

7. Gmelin, *Reise*.

8. K. A. Volosovich, *The Bolshoi Lyakhov Island Mammoth (New Siberia), a Geological Sketch* (in Russian) (Petrograd, 1915).

9. Peter Simon Pallas, *Commentaries of the St. Petersburg Academy for 1772*, 17:572.

10. Evert Ysbrants Ides, *Dreyjährige Reise nach China, von Moscou ab zu Lande durch gross-Ustiga, Sirianan, Permis, Sibirien, Daoum und die grosse Tartarey* (Frankfurt, 1707); trans. as *Three Years Travels from Moscow Overland to China* (London, 1707).

11. Gmelin, *Reise*.

12. Gmelin, *Reise*, 38.

13. Vassily Tatischev, "Generosiss. Dr Basilii Tatischow Epistola ad d. Ericum Benzelium de Mamontowa Kost, id est, de ossibus bestia Russis *Mamont* dicta," *Acta literaria Sveciae* (Stockholm and Uppsala, 1st trim., 1725).

14. Vassily Tatischev, *Skazanije o zvérié mamontié, o kotorom obyvatiéli sibirskijé ska-zaiout, iakoby jiviot pod zemliou, s ikh o tom dokazatielstvy i drougyikh o tom razlitchnyie mnienija* (Legends about the mammoth animal, which, according to the inhabitants of Siberia, lives underground, with their evidence and other opinions on the subject) (Upsalla, 1730); all translations of this work are mine.

15. Tatischev, *Skazanije o zvérié mamontié*.

16. Ibid.

17. Ibid.

18. Ibid.

19. Ibid.

20. Ibid.

21. Ides, *Dreyjährige Reise*.

22. Tatischev, *Skazanije o zvérié mamontié*.

23. Daniel Gottlieb Messerschmidt, quoted in Johann Philip Breyne, "Observations and a Description of Some Mammoth's Bones and Teeth Dug Up in Siberia, Proving Them to Have Belonged to Elephants," *Philosophical Transactions* (1737–38) (London: 1741).

24. Breyne, "Observations," 128.

25. Ibid.

26. Ibid.

27. Ibid.

28. Hans Sloane, *Philosophical Transactions* (1728) nos. 403–4.

29. Breyne, "Observations."

30. Ibid.

31. "Ivoire fossile," *Encyclopédie ou Dictionnaire raisonné des sciences, des arts et des métiers* (Société de gens de lettres, 1765), 9:64.

32. Breyne, "Observations," 129.

33. See chapter 3.

34. Thomas Burnet, *Telluris theoria sacra* (1681); trans. as *The Sacred Theory of the Earth* (London, 1684).

35. Johann Jakob Scheuchzer, *Physica sacra*, vols. 1–8 (Ulm, 1730–35); trans. as *Physique sacrée ou histoire naturelle de la Bible* (Amsterdam, 1732–37).

36. Scheuchzer, *Physique*, ii–iii.

37. Ibid., 65.

38. Ibid., 68.

39. Johann Jakob Scheuchzer, *Herbarium diluvianum* (Zurich, 1709).

40. Scheuchzer, *Physique*, 70.

41. "Ivoire," *Encyclopédie*, 9:64.

42. Elie Bertrand, *Dictionnaire universel des fossiles propres et des fossiles accidentels* (The Hague: 1763), 2:248–49. Bertrand was the first pastor of the French church in Bern.

43. Ibid.

44. On the appearance of ivory, see Gmelin, *Reise*, 2:147.

45. Georges-Louis Leclerc de Buffon, *Théorie de la Terre* (Paris: Imprimerie Royale, 1749), art. 3.

Chapter 5

1. Thomas Jefferson, *Notes on the State of Virginia* (Chapel Hill: University of North Carolina Press, 1955). Cf. Silvio A. Bedini, *Thomas Jefferson and American Vertebrate Paleontology* (Charlottesville: Virginia Div. of Mineral Resources Publication, 1985), 61:2. "This modest volume [is] now considered to be the most important scientific and political work produced in America before the end of the eighteenth century."

2. Jefferson, *Notes*, 53–54.

3. Ibid., 54.

4. Ibid., 43.

5. Georges Leclerc de Buffon, *Histoire naturelle générale et particulière . . .* (Paris: Imprimerie Royale, 1761), 9:104.

6. Quoted in Jefferson, *Notes*, 58.

7. Georges-Louis Leclerc de Buffon, *Histoire naturelle*, 15 vols. (Paris: Imprimerie Royale, 1749–67); *Supplément*, 7 vols. (Paris: Imprimerie Royale, 1774–89). See Pietro Corsi, *The Age of Lamarck* (Berkeley: University of California Press, 1988), 1–7.

8. See Jacques Roger, "Buffon, Jefferson et l'homme américain," *Bulletins et mémoires de la Société d'anthropologie de Paris*, n.s., 1, nos. 3–4 (1989): 57–66.

9. Jean Guettard, *Mémoires de l'Académie des sciences* (Paris, 1752), 360, pl. ii (1756).

10. According to George Gaylord Simpson in "The Beginnings of Vertebrate Pale-ontology in North America," *Proceedings of the American Philosophical Society* 86, no. 1 (September 1942): 130–88, Guettard was referring to the map published by Bellin in 1744, in which Longueuil's discovery is indicated by the phrase, "Endroit où on a trouvé des os d'éléphant en 1729" [actually 1739], *Carte de la Louisiane, cours du Mississippi et pais voisins, dédiée à M. le Comte de Maurepas, Ministre et Secrétaire d'Etat Commandeur des Ordres du Roy* (Paris, 1744).

11. "The appearance of these lines and dots are what is seen as the grain of the ivory. It is visible in all ivories, but is more or less noticeable in different tusks. Among those ivories whose grain is noticeable enough for them to be called 'coarse-grained,' some are called 'large-grained,' to distinguish them from those with fine grain." *Mémoires de l'Académie des sciences* (Paris, 1762).

12. On Croghan, see Simpson, "Beginnings," 141.

13. Ibid.

14. Georges-Louis Leclerc de Buffon, *Les Epoques de la nature*, ed. Jacques Roger (Paris: Editions du Muséum National d'Histoire Naturelle, 1962, 1988), n. 7.

15. Letter from Collinson to Buffon, quoted by Buffon in *Epoques*, n. 9.

16. William Hunter, paper presented to the London Royal Society on February 25, 1768.

17. Benjamin Franklin, quoted in Simpson, "Beginnings."

18. This is what Buffon affirms, in a late addition to *Théorie de la Terre:* "Several of these enormous bones, which I believed belonged to unknown animals, and which I supposed were of extinct species, nevertheless appeared, upon careful examination, to belong to the elephant and hippopotamus species: but in truth to elephants and hippopotamuses larger than existing ones. I know of only one extinct land animal,

whose molar teeth I had drawn, along with their dimensions; the other large teeth and large bones I was able to collect belonged to elephants and hippopotamuses."

19. Buffon, "Premier discours," *Epoques*.

20. Buffon, *Epoques*, 226.

21. Jefferson, *Notes*, 44.

22. Buffon, *Epoques*, 227n.

23. See Roger's preface to Buffon, *Epoques*, xliii.

24. See especially footnotes 7 and 9 in *Epoques de la nature*, which concern the American fossils and the correspondence with Collinson and Daubenton.

25. Gmelin, *Voyage en Sibérie*, chap. 58.

26. Buffon, "Premier discours," *Epoques*, 19.

27. Ibid., 22. (I use the original *Epoques* [1778] pagination.)

28. *Théorie de la Terre* was published in 1749 in the first volume of *Histoire naturelle*. Buffon, who was able to read a summary of *Protogaea* in the *Acta Eruditorium* (Leipzig, 1693) borrowed directly from Leibniz the idea that the earth was originally a molten heavenly body.

29. Buffon, *Epoques*, 26.

30. Ibid., 176.

31. Ibid., 178.

32. See chapter 2.

33. Buffon, *Epoques*, 27.

34. For Buffon, the appearance and disappearance of species could be explained by physical reasons. Indestructible "organic molecules" coalesce to form organized beings. So when the conditions are right, new species can appear in areas empty of animal life. "Reindeer and other animals that can only subsist in very cold climates appeared last. Who knows but what with the passage of time, when the earth will be colder, that new species may appear, whose temperament is as different from the reindeer as the reindeer in this respect is different from the elephant." *Epoques*, 176.

35. Buffon, "Premier discours," *Epoques*, 228n.

36. See the detailed presentation of these experiments in the *Supplément à l'Histoire naturelle*, vol. 1 (1774), experimental part, first and second paper; vol. 2 (1775) experimental part, eighth paper. For Roger's analysis of these texts and of the representation of time in *Epoques de la nature*, see pages xl–lxvii.

37. So Buffon had conceived the framework of his "system" of the *Epoques de la nature* as early as 1767. Could it have been profoundly influenced by his reading Dortous de Mairan's *Dissertation sur la glace* (Paris: Imprimerie Royale, 1749), as Roger suggests in his preface (p. xci)? But the *Dissertation* was published in 1749. One can imagine that Collinson's paper on the "unknown animal of the Ohio" and the questions he put to Buffon on the probable presence of elephants in northern countries were not unrelated to Buffon's own preoccupations at the time.

38. Roger has carefully studied Buffon's different measurements and the changes in the numbers that appear in the manuscript at various stages. See preface, *Epoques*, lxv.

39. Buffon, *Epoques*, 1.

40. Roger, preface, *Epoques*, lix–lx.

41. Ibid., cxlv.

42. See Condorcet, *Eloge de M. le comte de Buffon* (Paris, 1790).

43. Georges Cuvier, *Recherches sur les ossemens fossiles de quadrupèdes*, 1st ed. (Paris, 1812), 2:42.

44. Johann Friedrich Blumenbach, *Handbuch für Naturgeschichte* (Göttingen, 1779); trans. by R. T. Gore as *A Manual on the Elements of Natural History* (London, 1825). "On Petrifications," sec. 262, p. 407.

45. See Simpson's title, "The Beginnings of Vertebrate Paleontology in North America."

46. See Joseph Ellis, *After the Revolution: Profiles of Early American Culture* (New York: Norton, 1979).

47. Rembrandt Peale, *Account of the Skeleton of the Mammoth* (London, 1802), and *An Historical Disquisition on the Mammoth* (London, 1803).

48. See Ronald Rainger, *An Agenda for Antiquity* (Tuscaloosa: University of Alabama Press, 1991).

Chapter 6

1. Georges Cuvier, "Mémoire sur les espèces d'éléphans, tant vivantes que fossiles," *Magasin encyclopédique* (Paris, 1796), 3:440–45, and "Mémoire sur les espèces d'éléphans vivantes et fossiles," *Mémoires de l'Institut national des sciences et des arts. Sciences mathématiques et physiques* 2 (1799): 1–22. Translator's note: Most of the Cuvier excerpts are from Martin J. S. Rudwick's elegant translation, *Georges Cuvier, Fossil Bones, and Geological Catastrophes* (Chicago: University of Chicago Press, 1997), and are so indicated in the notes.

2. Pieter Camper, *Description anatomique d'un éléphant mâle*, published by his son Adrien Gilles Camper, with 20 plates (Paris: Jansen, 1802). This elephant, wrote Adrien Camper, had been "transported from Ceylon with the menagerie of HRH the Prince of Orange, who, knowing how important the dissection of such a rare quadruped would be to natural science, ordered that in case it died, it should immediately be sent to Mr. Camper."

3. Cuvier, "Mémoires" (1799), 12; Rudwick, *Georges Cuvier*, 19n.

4. Cuvier, "Mémoires" (1799), 14–15.

5. See chapter 3.

6. Blumenbach, *Handbuch für Naturgeschichte*; trans. by R. T. Gore as *A Manual on the Elements of Natural History* (London, 1825). The *Manual* was originally published at Göttingen in 1779, but the sixth edition (1799) first mentions the classification of petrifications into "doubtful" and "unknown."

7. Blumenbach, *Manual*, 403.

8. Ibid., 404, 406.

9. See chapter 3, p. 58.

10. Blumenbach, *Manual*, 404.

11. Georges Cuvier, *Discours sur les révolutions de la surface du globe* (Paris: Bourgeois, 1985), 94; Rudwick, *Georges Cuvier*, 216.

12. Georges Cuvier, *Mémoires de l'Institut national des sciences et des arts*, vol. 52 (Paris, 1801), 253.

13. The word was first used in French by the naturalist Henri Ducrotay de Blainville in *Journal de physique* 94:liv.

14. Cuvier, "Mémoires" (1799).

15. This is exactly what had been noted by the mathematician and astronomer Laplace. See *Exposition du système du monde*, 2:138.

16. Cuvier, "Mémoires" (1796), 3:444; Rudwick, *Georges Cuvier*, 24.

17. Blumenbach, *Manual*.

18. On "actualism in geology," see Reijer Hooykaas, *Natural Law and Divine Miracle: The Principle of Uniformity in Geology, Biology and Theology* (Leiden: E. J. Brill, 1963).

19. Cuvier, "Mémoires" (1799); Rudwick, *Georges Cuvier*, 93.

20. Cuvier, *Discours*, 74; Rudwick, *Georges Cuvier*, 226.

21. See Cuvier and Brongniart, *Essai sur la géographie minéralogique du Bassin de Paris* (Paris, 1808).

22. Cuvier and Brongniart, *Essai*, 1; Rudwick, *Georges Cuvier*, 183.

23. Cuvier and Brongniart, *Essai*, 97; Rudwick, *Georges Cuvier*, 217.

24. Cuvier and Brongniart, *Essai*, 99; Rudwick, *Georges Cuvier*, 219.

25. Cuvier and Brongniart, *Essai*, 32.

26. Ibid., 74; Rudwick, *Georges Cuvier*, 206.

27. Cuvier and Brongniart, *Essai*, 260–61.

28. Ibid., 31.

29. Ibid., 32; Rudwick, *Georges Cuvier*, 185.

30. Cuvier and Brongniart, *Essai*, 42; Rudwick, *Georges Cuvier*, 190.

31. Cuvier and Brongniart, *Essai*, 42.

32. Ibid., 239–73.

33. Ibid., 117; Rudwick, *Georges Cuvier*, 228–29.

34. Translation edited by Robert Jameson under the title *Essay on the Theory of the Earth* (Edinburgh, 1813).

35. Cuvier, *Discours*, 225; Rudwick, *Georges Cuvier*, 248.

36. Cuvier, *Discours*, 126; Rudwick, *Georges Cuvier*, 233.

37. See Dorinda Outram, *Georges Cuvier: Vocation, Science and Authority in Post-Revolutionary France* (Manchester: Manchester University Press, 1984), 146. It has often been said that as a Protestant, Cuvier found it easier to balance science and religion than a Catholic would. "Scripture should be open to free interpretation by individuals" and to "the habit of *libre examen*," writes Outram.

38. Ibid., 142.

39. Honoré de Balzac, *La Peau de chagrin*; trans. as *The Wild Ass's Skin* (London: Everyman's Library, 1967), 19.

40. Cuvier, *Discours*, 42–44.

41. Ibid., 43; Rudwick, *Georges Cuvier*, 190.

42. Cuvier, *Discours*, 42–43; Rudwick, *Georges Cuvier*, 190.

43. Cuvier, *Discours*, 33; Rudwick, *Georges Cuvier*, 185–86.

44. Cuvier, *Discours*, 31; Rudwick, *Georges Cuvier*, 183.

45. Cuvier, *Discours*, 33; Rudwick, *Georges Cuvier*, 185.

46. Outram, *Georges Cuvier*, 141–60.

47. Cuvier, *Discours*, 35; Rudwick, *Georges Cuvier*, 186.

48. Cuvier, *Discours*, 35; Rudwick, *Georges Cuvier*, 186.

49. Jacques Delille, *Les Trois règnes de la nature* (Paris, 1801); Félix Boitard, *Paris avant les hommes* (Paris, 1861).

50. Louis Figuier, *La Terre avant le Déluge* (Paris: Hachette, 1861; 6th ed. 1867); trans. as *The Earth before the Deluge*, ed. Henry Bristow (London: Chapman & Hall, 1867).

51. See Desmond, *Archetypes and Ancestors*, and *The Politics of Evolution*.

Chapter 7

1. Richard Owen, *A History of British Fossil Mammals and Birds* (London, 1846), 256. On Owen, see Desmond, *Archetypes and Ancestors*; and Rupke, *Richard Owen*.

2. Owen, *History*, 246.

3. William Buckland, *Reliquiae Diluvianae; or Observations on the Organic Remains*

contained in Caves, Fissures, and Diluvial Gravel, and on Other Geological Phenomena, Attesting to the Action of an Universal Deluge (London, 1823).

4. See chapter 3.

5. See the titles of Buckland's works, including *Geology and Mineralogy as Exhibiting the Power, Wisdom, and Goodness of God* (London, 1869–70).

6. See chapter 9 on Buckland's interpretation of the diluvial origin of the fossil elephants found in Alaska.

7. Charles C. Gillispie, *Genesis and Geology: A Study in the Relations of Scientific Thought, Natural Theology, and Social Opinion in Great Britain, 1790–1850* (Cambridge: Harvard University Press, 1951), 98.

8. Ibid., 120.

9. See Buffetaut, *A Short History of Palaeontology.*

10. Thomas H. Huxley, "Owen's Position in the History of Anatomical Science," in Rev. R. S. Owen, *The Life of Richard Owen* (London, 1894), 2:310.

11. Richard Owen, *On the Anatomy of Vertebrates,* 3 vols. (London, 1866–68).

12. Richard Owen, *On the Nature of Limbs* (London, 1849), 22.

13. Ibid., 86.

14. Owen, *Anatomy of Vertebrates,* 3:796.

15. Owen's presidential address on September 22, 1858, is in *Reports of the British Association for the Advancement of Science.*

16. Owen, *Anatomy of Vertebrates,* 3:808.

17. Owen, *Principes d'ostéologie comparée* (Paris, 1855), 427; W. Rodarmor's translation.

18. Owen, *On the Nature of Limbs,* 39.

19. "There are few examples of natural structures that manifest a more striking adaptation of a highly complex and beautiful structure to the exigencies of the animal endowed with it, than the grinding teeth of the Elephant. Thus the jaw is not encumbered with the whole weight of the massive tooth at once, but it is formed by degrees as it is required; the subdivision of the crown into a number of successive plates, and of the plates themselves into subcylindrical processes, presenting the conditions most favourable to progressive formation. But a more important advantage is gained by the subdivision of the grinder: each part is formed like a perfect tooth, having a body of dentine, a coat of enamel, and an outer investment of cement." Owen, *History,* 229.

20. "The formation [of the molar] begins with the summits of the anterior plate, and the rest are completed in succession: the tooth is gradually advanced in position as the growth proceeds, and its anterior plates are brought into use before the posterior ones are formed. When it cuts the gum, the cement is first rubbed off the digital summits; then the enameled cap is worn away and the central dentine exposed; next the digital processes are ground down to their common uniting base and . . . the upper margin of the plate with its border of enamel is exposed; finally, the transverse plates themselves are abraded to their common base of dentine, and a smooth and polished tract of that substance of greater or lesser extent is produced. When the tooth has been thus reduced to an uniform surface it becomes useless as an instrument for grinding the coarse vegetable substances on which the Elephant subsists, and is shed." Ibid., 228.

21. Cuvier, *Recherches sur les ossemens fossiles de quadrupèdes,* 3rd ed. (1825), vol. 1, chap. 2.

22. Owen, *History.*

23. Filippo Nesti, "Di alcune ossa fossili de Mammiferi che s'incontrano nel Val d'Arno," *Annali del Museo di Firenze* 1 (1808).

24. Owen, *History*, 232.

25. Ibid., 243–44.

26. For a biography of Falconer, see Charles Murchison, "Biographical Sketch," in Hugh Falconer, *Palaeontological Memoirs*, vol. 1 (London: Hardwicke, 1868).

27. Falconer, *Memoirs*, 1:27.

28. Ibid., 51.

29. Murchison, "Biographical Sketch," 27.

30. Ibid., 52–53.

31. Rudwick, *The Meaning of Fossils*, 208.

32. Falconer, "On the Species of Mastodon and Elephant Occurring in the Fossil State in Great Britain," *Quarterly Journal of the Geological Society* (November 1857), reprinted in *Memoirs*, 2:1–211.

33. Falconer, *Memoirs*, 1:24.

34. Falconer, "On the Species," *Memoirs*, 2:2.

35. Ibid., 77.

36. Ibid., 78.

37. Ibid., 6.

38. See Hugh Falconer, April 8, 1857, communication to the London Geological Society, published in the *Quarterly Journal of the Geological Society* (November 1857), reprinted in *Palaeontological Memoirs*, 2:79.

39. Falconer, *Memoirs*, 2:9.

40. Ibid.

41. See Martin J. S. Rudwick's review of Peter Bowler's "Fossils and Progress," *American Journal of Science* 278:95–96.

42. Charles Darwin, *On the Origin of Species* (London, 1859), facsimile (Cambridge: Harvard University Press, 1964), 301.

43. Ibid., 279–311.

44. Ibid., 281.

45. Falconer, "On the American Fossil Elephant of the Region Bordering the Gulf of Mexico (*E. columbi, Falc.*), with General Observations on the Living and Extinct Species," *Natural History Review* (January 1863), reprinted in *Memoirs*, 2:212–91.

46. Falconer, *Memoirs*, 2:251.

47. Ibid., 254.

48. See Desmond, *Archetypes and Ancestors*.

Chapter 8

1. J. H. Rosny Aîné, *La Guerre du feu* (1909), chap. 4, "The Pact between Man and Mammoth," in *Romans préhistoriques* (Paris: Laffont, 1985), 262, trans. by Harold Talbott as *Quest for Fire* (New York: Ballantine Books, 1982).

2. Sigmund Freud, "A Difficulty in the Path of Psycho-analysis," *Standard Edition of the Complete Psychological Works of Sigmund Freud*, 24 vols., trans. James Strachey (London: Hogarth Press, 1953–74), 17:137–44.

3. "Observations sur les ossements humains et les objets de fabrication humaine confondus avec des ossements de mammifè res appartenant à des espèces perdues, par M. Tournal fils, de Narbonne," in *Bulletin des sciences naturelles et de géologie* (1831): 250–51.

4. Charles Fraipont, "Les Hommes fossiles d'Engis," *Archives de l'Institut de paléontologie humaine* 16 (1936).

5. "Since I can guarantee that none of these objects was introduced after the fact, I attach great importance to their presence in the caves; because even if we had not found human bones, in conditions completely favorable to their being considered as belonging to the antediluvian epoch, the carved bones and worked flints would have given us proof of it." Philippe C. Schmerling, *Mémoires de la Société géologique française*, séance du 16 mars 1835, 173.

6. See Léon Aufrère, *Essai sur les premières découvertes et les origines de l'archéologie primitive (1811–1844)* (Paris: Staude, 1936).

7. Jacques Boucher de Perthes, *Les Antiquités celtiques et antédiluviennes* (Paris, 1847), 1:238.

8. On the Brixham dig, see J. Prestwich, *Excavations at Brixham Cave*, 482, 499–516. For a study of English paleontological activity in 1858–63, see A. Bowdoin Van Riper, *Men among the Mammoths: Victorian Science and the Discovery of Human Prehistory* (Chicago: University of Chicago Press, 1993).

9. See Claudine Cohen and Jean-Jacques Hublin, *Boucher de Perthes: Les Origines romantiques de la préhistoire* (Paris: Belin, 1989).

10. Charles Lyell, *The Geological Evidences of the Antiquity of Man* (London: J. Murray, 1863).

11. J. B. Lamarck, *Philosophie zoologique* (Paris, 1809), 1:349–50.

12. Thomas Henry Huxley, *Evidence as to Man's Place in Nature* (London: Williams & Norgate, 1863). Ernst Haeckel, *Anthropogenie oder Entwickelungsgeschichte des Menschen* (Leipzig, 1874); trans. as *The Evolution of Man* (New York: Appleton, 1879). Charles Darwin, *The Descent of Man and Selection in Relation to Sex* (London: Murray, 1871).

13. See Desmond, *Archetypes and Ancestors*.

14. Edouard Lartet, "Nouvelles recherches sur la coexistence de l'homme et des grands mammifères fossiles réputés caractéristiques de la dernière période géologique," *Annales des sciences naturelles*, 4th series, 15 (1861).

15. Edouard Lartet, "Sur des figures d'animaux gravées et sculptées et autres produits d'art et d'industrie rapportables aux temps primordiaux de la période humaine," *Revue archéologique* (1864).

16. John Lubbock, *Pre-historic Times* (London: William & Norgate, 1865), 2.

17. Gabriel de Mortillet, *Le Préhistorique, antiquité de l'homme* (Paris: Reinwald, 1880).

18. See Yves Coppens, Vincent J. Maglio, T. Madden, Michel Beden, "Proboscidea," in *Evolution of African Mammals*, ed. V. J. Maglio and H. B. Cooke (Cambridge: Harvard University Press, 1978).

19. Edouard Lartet and Henry Christy, *Reliquiae Aquitanicae*, ed. Thomas Rupert (London, 1875), 206.

20. Edouard Piette, *L'Art pendant l'âge du renne* (Paris: Masson, 1907).

21. Edouard Piette, *L'Epoque éburnéenne et les races humaines de la période glyptique* (Saint-Quentin, 1894), 5–6.

22. L. R. Nougier and R. Robert, *Rouffignac*, vol. 1: *Galerie Henri Breuil et Grand plafond* (Florence: Sansoni, 1959).

23. André Leroi-Gourhan, *La Préhistoire de l'art occidental* (Paris: Mazenod, 1965), trans. by Norbert Guterman as *The Art of Prehistoric Man in Western Europe* (London: Thames & Hudson, 1968). See also *Le Fil du temps* (Paris: Fayard, 1983).

24. R. Desbrosses and J. Koslowski, *Hommes et climats à l'âge du mammouth* (Paris: Masson, 1988).

25. Ivan S. Poliakov, *Antropologicheskaia poiezdka v tsentral'nuiu i vostotchnuiu Rossi-*

jou, Ispolniennaia po porucheniiu Imperatorskoi akademii nauk (Anthropological travels in central and eastern Russia) (Saint-Petersburg: Zapisok Akademii Nauk, 1880), 9–43.

26. Ibid., 23–24; my translation.

27. The "horizontal method" (literally, "wide area method").

28. Abramova, "Correlations," 333–42.

29. Francine David, the "Kostienki" article in André Leroi-Gourhan's *Dictionnaire de préhistoire* (Paris: PUF, 1988), 577.

30. Abramova, "Correlations," 334.

31. Ibid.

32. Piotr Petrovitch Efimenko, *Znatchénijé Jenchtchiny v Orinjiakskoujou Epokhou* (The meaning of woman in the Aurignacian era) (1931).

33. Abramova, "Correlations."

34. Ibid., 338–39.

35. Joseph Stalin, "Marxism and the Problems of Linguistics," *Pravda*, June 20, 1950.

36. I want to thank professor Mikhail Anikovitch, a prehistorian at the Institute of Material Cultures at the St. Petersburg Academy of Sciences, for giving me valuable documents and information about the history of Soviet archaeology of the period during my visit to Kostienki in July 1992.

37. P. P. Efimenko, *Kostienki I* (Moscow-Leningrad, 1958).

38. This approach would be repeated in France by the prehistorian André Leroi-Gourhan starting in 1949 in the Arcy-sur-Cure Cave, then in 1960 on the Pincevent training site, an open-air late Paleolithic site on the left bank of the Seine below Montereau. See Leroi-Gourhan, *Le Fil du temps*.

39. See Jan Jelínek, *The Pictorial Encyclopedia of the Evolution of Man*, trans. Helga Hanks (London: Hamlyn, 1975).

40. Kurt Lindner, *La Chasse préhistorique*, trans. Georges Montandon (Paris: Payot, 1941).

41. Ibid., 172.

42. Lewis Binford, *In Pursuit of the Past* (New York: Thames & Hudson, 1984), 74–75.

Chapter 9

1. Charles Darwin, *On the Origin of Species* (London, 1859); facsimile (Cambridge: Harvard University Press, 1964).

2. Ibid., chap. 4, 129.

3. Albert Gaudry, *Animaux fossiles et géologie de l'Attique, d'après les recherches faites en 1855–56 et en 1860 sous les auspices de l'Académie des sciences* (Paris: Savy, 1862–67).

4. Gaudry, "Sur l'éléphant de Durfort," in *Centenaire de la fondation du Muséum d'histoire naturelle*, a commemorative volume published by the Muséum's professors (Paris: Imprimerie Nationale, 1893).

5. See Y. Conry, *L'Introduction du darwinismse en France* (Paris: Vrin, 1974).

6. Gaudry, *Cours de paléontologie*, opening lecture (Muséum National d'Histoire Naturelle, 1873), 18.

7. Gaudry, *Animaux fossiles*.

8. Gaudry, *Essai de paléontologie philosophique* (Paris: Masson, 1896), 187–92.

9. Gaudry, "Durfort," 25.

10. Ibid., 24.

11. Gaudry, *Paléontologie philosophique*, 30.

12. Cf. the title of his three-volume work, *Les Enchaînements du monde animal* (Paris: Hachette, 1878–90).

13. Letter from Darwin to Gaudry, January 21, 1868, in *Correspondance*, 2:396.

14. See P. J. Bowler, *The Eclipse of Darwinism* (Baltimore: Johns Hopkins University Press, 1983).

15. Henry F. Osborn, *Proboscidaea* (New York: Museum of Natural History, 1936–44). On Osborn's work, see Rainger, *An Agenda for Antiquity.*

16. Henry Fairfield Osborn, *The Origin and Evolution of Life* (New York: Scribner's, 1917), 114n.

17. See Rainger, "The Continuation of Morphological Tradition," *Journal of the History of Biology* 14, no. 1 (spring 1981): 129–58.

18. Charles Depéret, *Les Transformations du monde animal* (Paris: Flammarion, 1907); trans. as *The Transformations of the Animal World* (London: K. Paul, 1909).

19. Depéret, *Transformations*, 109.

20. C. Depéret and L. Mayet, "Monographie des éléphants pliocènes d'Europe et de l'Afrique du Nord," *Annales de l'université de Lyon* 1, no. 42 (Lyon-Paris: Sciences, Médecines, 1923): 91–224.

21. Charles Andrews, *A Descriptive Catalogue of the Tertiary Vertebrata of El Faiyûm, Egypt* (London: British Museum, 1906), 36.

22. Osborn, *Proboscidaea.*

23. On evolution along parallel ladders, see Henry F. Osborn, *The Age of Mammals* (New York: Macmillan, 1910).

24. François Jacob, *La Logique du vivant* (Paris: Gallimard, 1970). Trans. as *The Logic of Life: A History of Heredity*, trans. Betty E. Spillmann (New York: Pantheon, 1973).

25. See Peter J. Bowler, *Evolution: The History of an Idea* (Berkeley: University of California Press, 1984); Jean Gayon, *Une Histoire de l'hypothèse de sélection naturelle* (Paris: Kime, 1992). Trans. as *Darwinism's Struggle for Survival: Heredity and the Hypothesis of Natural Selection*, trans. Matthew Cobb (Cambridge: Cambridge University Press, 1998).

26. Theodosius Dobzhansky, *Genetics and the Origin of Species* (New York: Columbia University Press, 1937), 12.

27. Ernst Mayr, "Some Thoughts on the History of the Evolutionary Synthesis," in *The Evolutionary Synthesis*, ed. Ernst Mayr and William Provine (Cambridge: Harvard University Press, 1980), 1–48.

28. G. L. Jepsen, E. Mayr, and G. G. Simpson, eds., *Genetics, Paleontology and Evolution* (Princeton: Princeton University Press, 1949).

29. George Gaylord Simpson, *Tempo and Mode in Evolution* (New York: Columbia University Press, 1944), 3.

30. Stephen Jay Gould, "G. G. Simpson, Paleontology and the Modern Synthesis," in *The Evolutionary Synthesis*, ed. Mayr and Provine, 158.

31. Simpson, *Tempo and Mode*, 202–6.

32. Alfred Romer, "Time Series and Trends in Animal Evolution," in *Genetics, Paleontology and Evolution*, ed. Jepsen, Mayr, and Simpson, 103–20.

33. George Gaylord Simpson, *The Major Features of Evolution* (New York: Columbia University Press, 1953), 116.

34. Simpson, "The Principles of Classification and a Classification of Mammals," *Bulletin of the American Museum of Natural History* 85 (1945): 1–350.

35. Erich Thenius, "Phylogenie des Mammalia: Stammesgeschichte der Saügetiere (einschliesslich der Hominiden)," *Handbuch der Zoologie* 8, no. 2 (1969), 2.

36. Romer, "Time Series," 103.

37. Niles Eldredge and Stephen Jay Gould, "Punctuated Equilibria: An Alternative to Phyletic Gradualism," in *Models in Paleobiology*, ed. Thomas J. M. Shopf (San Francisco: Freeman, Cooper, 1972), 82–105. Reprinted in Niles Eldredge, *Time Frame: The Evolution of Punctuated Equilibria* (Princeton: Princeton University Press, 1989).

38. Eldredge and Gould, "Punctuated Equilibria," 84.

39. Ibid.

40. Stephen Jay Gould and Richard Lewontin, "The Spandrels of San Marco and the Panglossian Paradigm," in *Proceedings of the Royal Society* B 205 (London, 1979): 581–98.

41. See Vincent J. Maglio, "Origin and Evolution of the Elephanditae," *Transactions of the American Philosophical Society*, n.s., 63 (1973): 1–149.

42. Adrian M. Lister, "Evolution and Paleoecology of Elephants and Their Relatives in Eurasia," in *The Proboscidea: Evolution and Paleoecology of Elephants and Their Relatives*, ed. Jeheskel Shoshani and Pascal Tassy (Oxford: Oxford University Press, 1996).

43. Ibid.

Chapter 10

1. "Gold! Gold! Gold! Gold!" *Seattle-Post Intelligencer*, July 17, 1897, 1.

2. Jack London, *The Call of the Wild* (New York: Macmillan, 1903).

3. L. M. Prindle, *The Yukon-Tanana Region of Alaska: Description of the Circle Quadrangle* (Washington, D.C.: GPO, U.S. Geological Survey 1960), 20.

4. Otto von Kotzebue, *A Voyage of Discovery into the South Sea and Beering's Straits* (1821), trans. H. E. Lloyd (Amsterdam: N. Israel, 1967), 1:219–20.

5. William Buckland, "On the Occurrence of the Remains of Elephants . . . ," appendix to Frederick W. Beechey, *Narrative of a Voyage to the Pacific and the Beering's Strait* (1831), 2:332.

6. W. H. Dall, "Extract from a Report to C. P. Patterson, Supt. Coast and Geodetic Survey," *American Journal of Science* 21 (1881): 106.

7. A. G. Maddren, "Smithsonian Exploration in Alaska in 1904 in Search of Mammoth and Other Fossil Remains," Smithsonian Misc. Collections, vol. 49, no. 1584 (Washington, D.C., 1905).

8. Charles W. Gilmore, "Smithsonian Exploration in Alaska in 1907 in Search of Pleistocene Fossil Vertebrates, Second Expedition," Smithsonian Misc. Collections, vol. 51 (Washington, D.C., 1908).

9. Ibid., 4.

10. See Troy L. Pewe, *Quaternary Geology of Alaska, Geological Survey Professional Paper 835* (Washington, D.C.: GPO, 1975).

11. Gilmore, "Smithsonian Exploration," 29.

12. Eugen W. Pfizenmayer, *Siberian Man and Mammoth*, trans. Muriel Simpson (London: Blacking, 1939), 185.

13. Gilmore, "Smithsonian Exploration," 28.

14. Pfizenmayer, *Siberian Man*, 86.

15. Ibid.

16. Ibid., 9.

17. Ibid., 85.

18. Ibid., 90.

19. Ibid., 103.

20. Ibid.

21. Ibid., 104.

22. Ibid., 105.

23. Otto Herz, *Naoutchnyi rezoultaty ekspeditsii snariajennoj Imperatorskoj Akademii Nauk dlia raskopka mamonta, Naidiennavo na rekié Berezovke v 1901 godu* (Results of the 1901 scientific expedition to retrieve the Berezovka mammoth), vols. 1, 2 (St. Petersburg, 1903), vol. 3 (Petrograd, 1914).

24. N. K. Vereshchagin, ed., *Magadanskii mamontjonok* [the baby mammoth of Magadan], *Mammuthus primigenius (Blumenbach)*, USSR Academy of Sciences (Leningrad: Nauka, 1981).

25. See Maglio, "Origin and Evolution of the Elephantidae."

26. Coppens et al., "Proboscidea," in *Evolution of African Mammals*, 357.

27. E. James Dixon, *Quest for the Origins of the First Americans* (Albuquerque: University of New Mexico Press, 1993).

28. Ibid.

Chapter 11

1. See V. E. Garutt, *Das Mammut:* Mammuthus primigenius *(Blumenbach)* (Wittenberg, 1964).

2. See V. E. Garutt, A. Gentry, and A. M. Lister, "*Mammuthus* Brookes, 1828 (*Mammalia, Proboscidea*); Proposed Conservation, and *Elephas primigenius* Blumenbach, 1799 (currently *Mammuthus primigenius*): Proposed Designation as the Type Species of *Mammuthus*, and Designation of a Neotype," *Bulletin of Zoological Nomenclature* 47 (1990): 38–44.

3. Blumenbach, *Manual*. See "On Petrifications," 403.

4. See chapter 3, p. 58.

5. The two pamphlets in this collection are *Des Unicornu fossilis oder gegrabenen Einhorn Welches in der Herrshafft Tonna gefunden worden Berfertiget von dem Collegio Medico in Gotha den 14 febr. 1696* and *De sceleto elephantino a celeberrimo Wilhelmo Tentzelio Historigraphio ducali saxonica, ubi quoque Testaceorum petrificationes defenditur . . .* (Urbini: Litteris Leonardi, 1697).

6. I owe many thanks to Stephen Jay Gould for telling me about this precious document and letting me take pictures of it.

7. "International Code of Zoological Nomenclature Adopted by the 35th International Congress of Zoology, London, July 1958," in Ernst Mayr, *Principles of Systematic Zoology* (New York: McGraw-Hill, 1969).

8. See chapter 5, p. 100.

9. See Simpson, "The Principles of Classification," 240; and Arthur T. Hopwood, "Fossil Proboscidaea from China," *Palaeontologica sinica* 9, ser. C, pt. 3 (1935): 1–108.

10. Pascal Tassy, *L'Arbre à remonter le temps* (Paris: Bourgeois, 1991), 21. For a critical review of Proboscidea classifications, see Tassy, "Phylogénie et classification des *Proboscidaea (Mammalia)*, historique et actualité," *Annales de paléontologie* 76, no. 3 (1990): 159–224.

11. See David Hull, *Science as a Process* (Chicago: University of Chicago Press, 1988), 118–30.

12. W. Hennig, *Phylogenetic Systematics* (Urbana: University of Illinois Press, 1966).

13. Tassy, "Phylogénie."

14. Ibid., 220.

15. *PAUP v.2.4* developed by Swofford in 1985 and *Hennig 86 v.1.5* by Farris in 1988.

16. "I leave to various future times, but not to all, my garden of forking paths," wrote Ts'ui Pên, the Chinese sage in Borges's short story. "*The Garden of Forking Paths* [explained the brilliant English scholar] was the chaotic novel itself. The phrase 'to future times, but not to all' suggested the image of bifurcating in time, not in space. In all fiction, when a man is faced with alternatives he chooses one at the expense of the others. In the almost unfathomable Ts'ui Pên, he chooses—simultaneously— all of them. He thus *creates* various futures, various times which start others that will in their turn branch out and bifurcate in other times." Jorge Luis Borges, "The Garden of Forking Paths" (1944), trans. Helen Temple and Ruthven Todd in *Ficciónes* (New York: Grove Press, 1962).

17. Tassy, "Phylogénie."

18. Ibid.

19. J. Shoshani, D. A. Walz, M. Goodman, J. M. Lowenstein, and W. Prychodko, "Protein and Anatomical Evidence of the Phylogenetic Position of *Mammuthus primigenius* among the *Elephantidae,*" *Acta Zoologica Fennica* 170 (1985): 238–40.

20. Ibid., 239–40.

21. Hull, *Science as a Process.*

22. Niles Eldredge, "Cladism and Common Sense," in *Phylogenetic Analysis and Paleontology,* ed. J. Cracraft and N. Eldredge (New York: Columbia University Press, 1979), 194–95.

23. See André Adoutte, *La Mémoire de la terre* (Paris: Seuil, 1992), 214. See also A. F. Wilson, "The Molecular Basis of Evolution," *Scientific American* 253, no. 4 (October 1985): 164–73.

24. Adoutte, *Mémoire,* 230.

25. Ibid., 215.

26. Vereshchagin, *Magadanskii Mamontjonok.*

27. *Nature* 291 (1981): 409. See also J. Lowenstein, "Radio Immune Assay of Mammoth Tissue," *Acta Zoologica Fennica* 170 (1985): 233–35.

28. Shoshani et al., "Protein and Anatomical Evidence," 240.

29. See M. Noro, R. Masuda, I. A. Dubrovo, M. C. Yoshida, and M. Kato, "Phylogenetic Inference of the Woolly Mammoth *Mammuthus primigenius,* Based on Complete Sequences of Mitochondrial Cytochrome *b* and 12S Ribosomal RNA Genes," *Journal of Molecular Evolution* 46 (1998): 314–24. See also T. Osawa, S. Hashaui, and V. M. Mikhelson, "Phylogenetic Position of Mammoth and Steller's Sea Cow within *Tethytheria* Demonstrated by Mitochondrial DNA Sequences," *Journal of Molecular Evolution* 44 (1997): 406–13.

30. Debruyne, *Phylogénie moléculaire des Elephants.* See also V. Barriel, R. Debruye, and P. Tassy in *Journal of Molecular Evolution* (forthcoming).

31. P. H. Johnson, C. B. Olson, and M. Goodman, "Prospects for the Molecular Reconstruction of the Woolly Mammoth's Evolutionary History: Isolation and Characterization of Deoxyribonucleic Acid from the Tissue of *Mammuthus primigenius,*" *Acta Zoologica Fennica* 170 (1985): 225–31.

32. Régis Debruyne, personal communication.

33. See Buigues and Latreille, *Mammouth.*

34. Michael Benton, "Palaeomolecular Biology and the Relationships of the Mammoth," *Geology Today* 2, no. 5 (September 1986): 135–36.

35. See Mayor, *The First Fossil Hunters,* 249.

Chapter 12

1. On Burian, see chapter 1, note 14.

2. George F. Wright, *Asiatic Russia* (New York: McClure, 1903), 581.

3. See especially Augusta and Burian, *A Book of Mammoths*.

4. Palissy, *Discours admirable des eaux et des fontaines*.

5. Leibniz, Hanover ms. LH 37, 4, Ff. 14–15. "Stone shells of several unknown species are found that one would seek in vain in the sea, a sign that they are sports of nature, unless one holds that they are vanished species, which is not likely."

6. Buffon, *Histoire naturelle générale et particulière* . . . (1749), 2:18–41; *Histoire naturelle* (1766), 14:311–74; see also Buffon, *Oeuvres philosophiques*, ed. Jean Piveteau (Paris: PUF, 1954), 243, 396.

7. See chapter 5.

8. Alcide d'Orbigny, *Cours élémentaire de paléontologie et de géologie stratigraphiques* (Paris: Masson, 1849–52).

9. See Gillispie, *Genesis and Geology*.

10. Howorth, *The Mammoth and the Flood*.

11. Louis Agassiz, speech given at the opening session of the Société helvétique des sciences naturelles, published in *Actes de la Société helvétique des sciences naturelles*, 22nd session (Neuchâtel, 1837).

12. Elizabeth Cary-Agassiz, *Louis Agassiz: His Life and Correspondence*, vol. 1 (Boston, 1886), 289.

13. Ibid., 263–64.

14. Ibid., 296.

15. Louis Agassiz, *Geological Sketches*, second series (Boston, 1876), 77.

16. Agassiz, *Geological Sketches* (Boston, 1866), 208.

17. William R. Farrand, "Frozen Mammoths and Modern Geology," *Science* 133 (March 17, 1961): 729–35.

18. See for example Jody Dillow, "The Catastrophic Deep-Freeze of the Berezovka Mammoth," *Creation Research Society Quarterly* 14 (June 1977): 5–13.

19. For a more complete discussion of Lyell's intellectual journey regarding extinctions, see Donald K. Grayson, "Nineteenth Century Explanations," in *Quaternary Extinctions*, ed. Paul S. Martin and Richard G. Klein. (Tucson: University of Arizona Press, 1984), 5–39.

20. Charles Lyell, *Principles of Geology* (London, 1830; facsimile, Berkeley: University of California Press, 1990).

21. Ibid., 1:97.

22. The phrase was coined by the geologist and historian of science William Whewell around 1832.

23. Lartet and Christy, *Reliquiae Aquitanicae*.

24. Lyell, *The Geological Evidences*, 374.

25. See Van Riper, *Man among the Mammoths*, 177–78.

26. Owen, *History of British Fossil Mammals and Birds*, 270.

27. Gaudry, *Essai de paléontologie philosophique*, 43.

28. See Rainger, *Agenda for Antiquity*, 123.

29. I. P. Tolmachoff, "The Carcasses of the Mammoth and Rhinoceros Found in the Frozen Ground of Siberia," *Transactions of the American Philosophical Society* 23 (1929).

30. Henry Neuville, "De l'extinction du mammouth," *L'Anthropologie* 29 (1918–19): 193–212.

31. Tolmachoff, "Carcasses," 65.

32. Ibid., 70.

33. Madison Grant, *The Passing of the Great Race*, preface by H. F. Osborn (London, 1920).

34. Darwin, *Origin of Species*, 109.

35. For a discussion (and criticism) of the "Darwinian paradigm" of extinctions, see David M. Raup, *Extinction: Bad Genes or Bad Luck?* (New York: Norton, 1991).

36. Darwin, *Origin of Species*, 322.

37. Romer, "Time Series," 118.

38. Ibid., 119.

39. Darwin, *Origin of Species*, 317–18.

40. See Raup, *Extinction*. Stephen Jay Gould develops the same thesis regarding Cambrian extinctions in *Wonderful Life* and *The Flamingo's Smile*.

41. Raup, *Extinction*, 5, 191.

42. Andrei V. Sher, *Pleistocene Mammals and Stratigraphy of the Far-East USSR and North America* (in Russian) (Moscow: Geological Institute, Academy of Sciences, 1971); trans. in *International Geological Review* 16 (1974): 1–224.

43. R. Dale Guthrie, *Frozen Fauna of the Mammoth Steppe* (Chicago: University of Chicago Press, 1990).

44. N. K. Vereshchagin and G. F. Barychnikov, "Quaternary Mammalian Extinctions in Northern Eurasia," in *Quaternary Extinctions*, ed. Martin and Klein, 483–516.

45. George Bassett Digby, *The Mammoth and Mammoth-hunting in North-east Siberia* (London: H. F. & G. Witherby, 1926).

46. Paul S. Martin, "Prehistoric Overkill," in *Pleistocene Extinctions: The Search for a Cause*, ed. P. S. Martin and H. E. Wright Jr. (New Haven: Yale University Press, 1967), 75–120.

47. Donald Grayson, "Explaining Pleistocene Extinctions: Thoughts on the Structure of a Debate," in *Quaternary Extinctions*, ed. Martin and Klein, 807–23.

48. Paul S. Martin, "Prehistoric Overkill: The Global Model," in *Quaternary Extinctions*, ed. Martin and Klein, 370.

49. See Desbrosses and Koslowski, *Hommes et climats à l'âge du mammouth*.

50. Vereshchagin and Barychnikov, "Mammalian Extinctions." See also N. K. Vereshchagin, *Pochemu vymerli mamonty* (Why mammoths became extinct) (Leningrad: Nauka, 1979).

51. S. L. Vartanyan, V. E. Garutt, and A. V. Sher, "Holocene Dwarf Mammoths from Wrangel Island in the Siberian Arctic," *Nature* 362 (March 25, 1993): 337–40.

52. A. M. Lister, "Mammoths in Miniature," *Nature* 362 (March 25, 1993): 288–89.

53. Vartanyan et al., "Holocene," 340.

54. *Elephas falconeri*—at less than three feet tall, a miniature elephant and the smallest of the Elephantidae family—also lived on islands, in Malta and Sicily. It became extinct in Roman times.

55. Vartanyan et al., "Holocene," 339.

56. See Jean-Pierre Cavaillé, *Descartes, la fable du monde* (Paris: Vrin-EHESS, 1991).

Conclusion

1. See Rudwick, *The Meaning of Fossils*, 266–67.

On the History of Paleontology

Adams, Frank Dawson. *The Birth and Development of the Geological Sciences.* New York: Dover Publications, 1938.

Appel, Toby A. *The Cuvier-Geoffroy Debate. French Biology in the Decades before Darwin.* Oxford: Oxford University Press, 1987.

Bedini, Silvio A. *Thomas Jefferson and American Vertebrate Paleontology.* Charlottesville: Virginia Division of Mineral Resources Publication, 61, 1985.

Buffetaut, Eric. *Des fossiles et des hommes.* Paris: Laffont, 1991. Trans. of *A Short History of Vertebrate Palaeontology.* London: Croom Helm, 1987.

Cohen, Claudine. "André Leroi-Gourhan, chasseur de préhistoire." *Critique* 444 (May 1984): 384–403.

———. "Formes et métamorphoses du récit paléontologique." In "Le Narratif hors fiction," ed. R. le Huenen. *Texte, Revue de critique et de théorie littéraire,* nos. 19–20 (Toronto, 1996): 279–89.

———. *La Genèse de* Telliamed. *Benoît de Maillet et l'histoire naturelle à l'aube des Lumières.* Thèse de l'université de Paris III, 1989, à paraître, Paris, Vrin-EHESS.

———. *L'Homme des origines: Savoirs et fictions en préhistoire.* Paris: Seuil, 1999.

———. "Richard Owen: Paléontologie, embryologie et morphologie transcendantale vers 1840." In *Actes du colloque "Les Philosophies de la nature,"* ed. O. Bloch. Paris: Presses de la Sorbonne, 2000.

Cohen, Claudine, and Jean-Jacques Hublin. *Boucher de Perthes: Les Origines romantiques de la préhistoire.* Paris: Belin, 1989.

Coleman, W. *Georges Cuvier Zoologist: A Study in the History of Evolution Theory.* Cambridge: Harvard University Press, 1964.

Desmond, Adrian. *Archetypes and Ancestors: Paleontology in Victorian London, 1850–1875.* Chicago: University of Chicago Press, 1982.

———. *The Politics of Evolution: Morphology, Medicine, and Reform in Radical London.* Chicago: University of Chicago Press, 1989.

Ellenberger, François. *Histoire de la géologie.* Vols. 1, 2. Paris: Lavoisier, 1994.

Gohau, Gabriel. *Une Histoire de la géologie.* Paris: Seuil, 1990. Trans. by Albert Carozzi and Marguerite Carozzi as *A History of Geology.* New Brunswick, N.J.: Rutgers University Press, 1991.

———. *Les Sciences de la terre aux XVIIᵉ et XVIIIᵉ siècles. Naissance de la géologie.* Paris: Albin Michel, 1990.

Gould, Stephen Jay. *Time's Arrow and Time's Cycle: Myth and Metaphor in the Discovery of Geological Time.* Cambridge: Harvard University Press, 1987.

Gould, Stephen Jay, and Rosamond Purcell. *Finders, Keepers: Eight Collectors*. New York: Norton, 1992.

Grayson, Donald. *The Establishment of Human Antiquity*. New York: Academic Press, 1983.

Greene, John. *The Death of Adam: Evolution and Its Impact on Western Thought*. Ames: Iowa State University Press, 1959.

Hölder, Helmut. *Kurze Geschichte der Geologie und Paläontologie*. Berlin: Springer-Verlag, 1989.

Hooykaas, R. *Continuité et discontinuité en géologie et en biologie*. Paris: Seuil, 1970.

Hull, David. *Science as a Process*. Chicago: University of Chicago Press, 1988.

Jacob, François. *La Logique du vivant: Une Histoire de l'hérédité*. Paris: Gallimard, 1970. Trans. as *The Logic of Life: A History of Heredity*, trans. Betty E. Spillmann. New York: Pantheon, 1973.

Laurent, Goulven. *Paléontologie et évolution en France, 1800–1860, de Cuvier-Lamarck à Darwin*. Paris: Editions du Comité des travaux historiques et scientifiques, 1987.

Mayor, Adrienne. *The First Fossil Hunters: Paleontology in Greek and Roman Times*. Princeton: Princeton University Press, 2000.

Mayr, E., and W. Provine. *The Evolutionary Synthesis: Perspectives on the Unification of Biology*. Cambridge: Harvard University Press, 1980.

Morello, Nicoletta. *La Nascita della paleontologia nel seicento. Colonna, Stenone e Scilla*. Ed. Franco Angeli. Milano, 1979.

Oldroyd, David R. *Thinking about the Earth: A History of Ideas in Geology*. London: Athlone, 1996.

Olmi, Giuseppe. *L'inventario del mondo: Catalogazione della natura e luoghi del sapere nella prima età moderna*. Bologna: Società editrice il Mulino, 1992.

Outram, Dorinda. *Georges Cuvier: Vocation, Science and Authority in Post-Revolutionary France*. Manchester: Manchester University Press, 1984.

Rainger, Ronald. *An Agenda for Antiquity: Henry Fairfield Osborn and Vertebrate Paleontology at the American Museum of Natural History, 1890–1935*. Tuscaloosa: University of Alabama Press, 1991.

Roger, Jacques, ed. *Buffon*. Paris: Fayard, 1989.

———. *Epoques de la nature*. Paris: Editions du Muséum d'histoire naturelle, 1962, 1988.

Rossi, Paolo. *I segni del tempo: Storia della terra e storia delle nazioni da Hooke a Vico*. Milan: Feltrinelli, 1979. Trans. by Lydia G. Cochrane as *The Dark Abyss of Time: The History of the Earth and the History of Nations from Hooke to Vico*. Chicago: University of Chicago Press, 1984.

Rudwick, Martin J. S. *Georges Cuvier, Fossil Bones, and Geological Catastrophes: New Translations and Interpretations of the Primary Texts*. Chicago: University of Chicago Press, 1997.

———. *The Great Devonian Controversy: The Shaping of Scientific Knowledge among Gentlemanly Specialists*. Chicago: University of Chicago Press, 1985.

———. *The Meaning of Fossils: Episodes in the History of Paleontology*. 2nd ed. Chicago: University of Chicago Press, 1976.

———. *Scenes from Deep Time: Early Pictorial Representations of the Prehistoric World*. Chicago: University of Chicago Press, 1992.

Rupke, Nicolaas. *Richard Owen: Victorian Naturalist*. New Haven: Yale University Press, 1994.

Shepard, Odell. *The Lore of the Unicorn.* London: Shepard, Allen & Unwin, 1930.

Simpson, George Gaylord. "The Beginnings of Vertebrate Paleontology in North America." *Proceedings of the American Philosophical Society* 86, no. 1 (September 1942): 130–88.

Tassy, Pascal. *L'Arbre à remonter le temps.* Paris: Bourgois, 1991.

Whewell, William. *History of the Inductive Sciences.* 2 vols. London: John W. Parker, 1837.

Zittel, Karl von. "Geschichte der Geologie und Paläeontologie bis Ende des 19 Jahrhunderts." *Geschichte der Wissenschaft in Deutschland.* Vol. 23. Munich: Oldenbourg, 1899.

Monographs on the Mammoth

Auel, Jean. *The Mammoth Hunters.* New York: Bantam, 1986.

Augusta, Josef, and Zdenek Burian. *Das Buch von den Mammuten.* Prague: Artia, 1962. Trans. by Margaret Schierl as *A Book of Mammoths.* London: Hamlyn, 1963.

Benton, Michael. "Palaeomolecular Biology and the Relationships of the Mammoth." *Geology Today* 2, no. 5 (September 1986).

Breyne, Johann Philip. "Observations, and a Description of Some Mammoth's Bones Dug Up in Siberia, Proving Them to Have Belonged to Elephants." *Philosophical Transactions* 40 (1737–38); London, 1741, 124–39.

Cohen, Claudine. "Sto liet (1695–1796) v istorii izoutchénija mamonta: nakhodki mamontovykh kostiei i ikh obiasnienija v zapadnoj Evropié" (A hundred years of studies on the woolly mammoth). *Troudy Zoologuitcheskovo Institouta, Rosiiskaia Akademiia Naouk* (Publications of the Zoological Institute of Saint Petersburg, Russian Academy of Science) 270 (1996): 196–205.

Coppens, Yves, Vincent J. Maglio, T. Madden, and Michel Beden. "Proboscidaea." In *Evolution of African Mammals,* ed. Vincent J. Maglio and H. B. Cooke. Cambridge: Harvard University Press, 1978.

Cuvier, Georges. "Mémoire sur les espèces d'éléphans, tant vivantes que fossiles." *Magasin encyclopédique* 3. Paris, 1796.

———. "Mémoire sur les espèces d'éléphans vivantes et fossiles." *Mémoires de l'Institut national des sciences et des arts. Sciences mathématiques et physiques* 2 (1799): 1–22.

Depéret, Charles, and Lucien Mayet. "Monographie des éléphants pliocènes d'Europe et de l'Afrique du Nord." *Annales de l'université de Lyon* 1, no. 42 (Lyon-Paris: Sciences, Médecines, 1923): 91–224.

Desbrosses, René, and Janusz Koslowski. *Hommes et climats à l'âge du mammouth.* Paris: Masson, 1988.

Digby, George Bassett. *The Mammoth and Mammoth-hunting in North-east Siberia.* London: H. F. & G. Witherby, 1926.

Dillow, Jody. "The Catastrophic Deep-Freeze of the Berezovka Mammoth." *Creation Research Society Quarterly* 14 (June 1977): 5–13.

Escutenaire, Catherine, Janusz K. Kozlowski, Valery Sitlivy, and Krzysztof Sobczyk. *Les Chasseurs de mammouths de la vallée de la Vistule.* Krakow-Spadzista B, *un site gravettien à amas d'ossements de mammouths.* Bruxelles: Musées royaux d'art et d'histoire et Université Jagellon de Cracovie, 1999.

Falconer, Hugh. *Palaeontological Memoirs.* Vol. 2, ed. by Charles Murchison. London: Hardwicke, 1868.

Farrand, William R. "Frozen Mammoths and Modern Geology." *Science* 133 (March 17, 1961): 729–35.

Garutt, Vadim E. *Das Mammut: Mammuthus primigenius (Blumenbach)*. Wittenberg, 1964.

Garutt, Vadim E., A. Gentry, and Adrian M. Lister. "*Mammuthus* Brookes, 1828 (*Mammalia, Proboscidea*) Proposed Conservation, and *Elephas primigenius* Blumenbach, 1799 (currently *Mammuthus primigenius*) Proposed Designation as the Type Species of *Mammuthus*, and Designation of a Neotype." *Bulletin of Zoological Nomenclature* 47 (1990): 38–44.

Gherbrant, Emmauel, Jean Sudre, and Henri Capetta. "A Palaeocene Proboscidean from Morocco." *Nature* 383 (September 1996).

Ginsburg, Léonard. "Nouvelles lumières sur les ossements fossiles autrefois attribués au géant Theutobochus." *Annales de paléontologie* 70 (1984): 181–219.

Guthrie, R. Dale. *Frozen Fauna of the Mammoth Steppe: The Story of Blue Babe*. Chicago: University of Chicago Press, 1990.

Haynes, Gary. *Mammoths, Mastodons and Elephants: Biology, Behaviour and the Fossil Record*. New York: Columbia University Press, 1991.

Herz, Otto. *Naoutchnyi rezoultaty ekspeditsii snariajennoj Imperatorskoi Akademii Nauk dlia raskopka mamonta, Naidiennavo na rekié Berezovke v 1901 godou* (Results of the 1901 scientific expedition to retrieve the Berezovka mammoth). Vols. 1, 2, Saint-Petersburg, 1903; vol. 3, Petrograd, 1914.

Hopwood, Arthur T. "Fossil Proboscidaea from China." *Palaeontologica sinica* 9 ser. C, pt. 3 (1935): 1–108.

Howorth, Henry H. *The Mammoth and the Flood*. London, 1887.

Ivanov, S. V. "The Mammoth in the Art of the Peoples of Siberia" (in Russian). *Publications of the Anthropological and Ethnographical Museum of Saint-Petersburg* XI (1949): 133–61.

Jelinek, Jan. *Encyclopédie illustrée de l'homme préhistorique*. Prague: Artia, 1974; Paris: Gründ, 1979.

Johnson, P. H., C. B. Olson, and M. Goodman. "Prospects for the Molecular Biological Reconstruction of the Woolly Mammoth's Evolutionary History Isolation and Characterization of Deoxyribonucleic Acid from the Tissue of *Mammuthus primigenius*." *Acta Zoologica Fennica* 170 (1985): 225–31.

Lister, Adrian M. "Mammoths in Miniature." *Nature* 362 (March 25, 1993): 288–89.

Lister, Adrian M., and Paul Bahn. *Mammoths*. New York: Macmillan, 1994.

Lowenstein, Jerold M. "Radio Immune Assay of Mammoth Tissue." *Acta Zoologica Fennica* 170 (1985): 233–35.

Maddren, A. G. "Smithsonian Exploration in Alaska in 1904 in Search of Mammoth and Other Fossil Remains." Smithsonian Miscellaneous Collections, vol. 49, no. 1584. Washington, D.C., 1905.

Maglio, Vincent J. "Origin and Evolution of the Elephantidae." *Transactions of the American Philosophical Society of Philadelphia*, n.s., 63 (1973): 1–149.

Merck, K. *Lettre sur les os fossiles d'éléphants et de rhinocéros qui se trouvent en Allemagne*. Darmstadt, 1783.

Nesti, Filippo. "Di alcune ossa fossili de Mammiferi che s'incontrano nel Val d'Arno." *Annali del Museo di Firenze* 1 (1808).

Neuville, Henry. "De l'extinction du mammouth." *L'Anthropologie* 29 (1918–19): 193–212.

Noro, M., R. Masuda, I. A. Dubrovo, M. C. Yoshida, and M. Kato. "Phylogenetic Inference of the Woolly *Mammoth Mammuthus primigenius*, Based on Complete

Sequences of Mitochondrial Cytochrome *b* and 12S Ribosomal RNA Genes." *Journal of Molecular Evolution* 46 (1998).

Nougier, Louis-René, and R. Robert. *Rouffignac.* Vol. 1, *Galerie Henri Breuil et Grand plafond.* Florence: Sansoni, 1959.

Osawa, T., S. Hashaui, and V. M. Mikhelson. "Phylogenetic Position of Mammoth and Steller's Sea Cow within *Tethytheria* Demonstrated by Mitochondrial DNA Sequences." *Journal of Molecular Evolution* 44 (1997).

Osborn, Henry F. *Proboscidaea.* 2 vols. New York: Museum of Natural History, 1936–44.

Peale, Rembrandt. *Account of the Skeleton of the Mammoth.* London, 1802.

———. *An Historical Disquisition on the Mammoth.* London, 1803.

Pfizenmayer, Eugen W. *Siberian Man and Mammoth,* trans. Muriel Simpson. London: Blacking, 1939.

Sher, Andrei V. *Pleistocene Mammals and Stratigraphy of Far-East USSR and North America* (in Russian). Moscow: Geological Institute, Academy of Sciences, 1971; trans. in *International Geological Review,* no. 16 (1974): 1–224.

Shoshani, Jeheskel, and Pascal Tassy, eds. *The Proboscidea: Evolution and Palaeoecology of Elephants and Their Relatives.* Oxford: Oxford University Press, 1996.

Shoshani, J., D. A. Walz, M. Goodman, J. M. Lowenstein, and W. Prychodko. "Protein and Anatomical Evidence of the Phylogenetic Position of *Mammuthus primigenius* among the *Elephantidae.*" *Acta Zoologica Fennica* 170 (1985): 237–40.

Surmely, Frédéric. *Le Mammouth, géant de la préhistoire.* Paris: Solar, 1993.

Tassy, Pascal. "Phylogénie et classification des *Proboscidaea (Mammalia),* historique et actualité." *Annales de paléontologie* 76, no. 3 (1990): 159–224.

Tatischev, Vassily. "Generosiss. Dr Basilii Tatischow Epistola ad d. Ericum Benzelium de Mamontowa Kost, id est, de ossibus bestia Russis *Mamont* dicta." *Acta literaria Sveciae* (Stockholm and Upsalla, 1725).

———. *Skazanije o zvérié mamontié, o kotorom obyvatiéli sibirskijé skazaiout, iakoby jiviot pod zemliou, s ikh o tom dokazatielstvy i drougyikh o tom razlitchnyie mnienija* (Legends about the animal mammoth, which, according to the inhabitants of Siberia, lives underground, with their evidence and other opinions on the subject). Upsalla, 1730.

Tentzelius, Wilhelm. *De sceleto elephantino a celeberrimo Wilhelmo Tentzelio Historiographo ducali saxonico, ubi quoque Testaceorum petrificationes defenduntur . . .* Urbini: Litteris Leonardi, 1697.

Tolmachoff, I. P. "The Carcasses of the Mammoth and Rhinoceros Found in the Frozen Ground of Siberia." *Transactions of the American Philosophical Society* 23 (1929).

Van Riper, A. Bowdouin. *Men among the Mammoths: Victorian Science and the Discovery of Human Prehistory.* Chicago: University of Chicago Press, 1993.

Vartanyan, S. L., V. E. Garutt, and A. V. Sher. "Holocene Dwarf Mammoths from Wrangel Island in the Siberian Arctic." *Nature* 362 (March 25, 1993): 337–40.

Vaufrey, Raymond. *Les Eléphants nains des îles méditerranéennes et la question des isthmes pléistocènes.* Paris: Masson, 1929.

Vereshchagin, N. K. *Pochemu vymerli mamonty* (Why mammoths became extinct). Leningrad: Nauka, 1979.

———, ed. *Magadanskii mamontjonok* [the baby mammoth of Magadan], *Mammuthus primigenius (Blumenbach).* USSR Academy of Science. Leningrad: Nauka, 1981.

Volosovich, K. A. *The Bolshoi Lyakhov Mammoth (New Siberia), A Geological Sketch* (in Russian). Petrograd, 1915.

Other Works

Abel, Othenio. *Geschichte und Methode der Rekonstruktion vorzeitlicher Wirbeltiere.* Jena: Gustav Fischer, 1925.

———. *Vorzeitliche Tierreste im Deutschen Mythus, Brauchtum und Volksglauben.* Jena: Gustav Fischer, 1939.

Agassiz, Louis. "Discours prononcé à l'ouverture des séances de la Société helvétique des sciences naturelles le 24 July 1837." *Actes de la Société helvétique des sciences naturelles,* 22nd Session. Neuchâtel, 1837.

———. *An Essay on Classification.* London, 1859.

———. *Geological Sketches.* Boston, 1866; second series, Boston, 1876.

Agassiz, Louis, A. Guyot, and E. Desor. *Système glaciaire ou recherches sur les glaciers, leur mécanisme, leur ancienne extension et le rôle qu'ils ont joué dans l'histoire de la terre.* Paris: Masson, 1847.

Aldrovandi, Ulisse. *De quadrupedibus solipedibus.* Bologna, 1639.

———. *Museum metallicum, in libros IV distributum,* ed. Bartholomeo Ambrosinus. Bologna, 1648.

Augustine of Hippo. *De Civitate Dei.* In *Basic Writings of Saint Augustine,* ed. Whitney J. Oates. New York: Random House, 1948.

Balzac, Honoré de. *La Peau de chagrin, Roman philosophique.* The Hague, 1831. Trans. as *The Wild Ass's Skin.* London: Everyman's Library, 1967.

Bauer, Georg (Agricola). *Bermannus, sive de Re Metallica.* Basel, 1530.

———. *De natura fossilium.* Basel, 1536.

Beechey, Captain Frederick W. *Narrative of a Voyage to the Pacific and the Beering's Strait, Part II, 1831, with an Appendix "On the Occurrence of the Remains of Elephants, and Other Quadrupeds, in the Cliffs of Frozen Mud, in Eschscholtz Bay, within Beering's Strait and in Other Distant Parts of the Shores of the Arctic Sea, by the Rev. W. Buckland [. . .] professor of geology and mineralogy in the University of Oxford.*

Belaval, Yvon. *Leibnitz, initiation à sa philosophie.* Paris: Vrin, 1975.

Bertrand, Elie. *Dictionnaire universel des fossiles propres et des fossiles accidentels.* 2 vols. The Hague, 1763.

Binford, Lewis. *In Pursuit of the Past.* London: Thames & Hudson, 1984.

Blumenbach, Johann Friedrich. *Handbuch für Naturgeschichte.* 1st German ed. Göttingen, 1779. Trans. by R. T. Gore as *A Manual on the Elements of Natural History.* London, 1825.

Blundell, Derek J., and A. C. Scott, eds. *Lyell: The Past Is the Key to the Present.* London: Geological Society Special Publications, 1998.

Boccaccio. *De genealogia deorum gentilium.* Book 15. 1481.

Boitard, Félix. *Paris avant les hommes.* Paris, 1861.

Borges, Jorge Luis. "The Garden of Forking Paths." *Ficciónes,* trans. Helen Temple and Ruthven Todd. New York: Grove Press, 1962.

Boucher de Perthes, Jacques. *Les Antiquités celtiques et antédiluviennes.* Vol. 1. Paris, 1847; vol. 2, Paris, 1857; vol. 3, Paris, 1864. Reprint, Paris: Jean-Michel Place, 1989.

Brookes, Joshua. *Catalogue of the Anatomical & Zoological Museum of Joshua Brookes.* London: R. Taylor, 1828.

Buckland, William. *Reliquiae Diluvianae; or Observations on the Organic Remains Contained in Caves, Fissures, and Diluvial Gravel, and on Other Geological Phenomena, Attesting to the Action of an Universal Deluge.* London, 1823.

Buffon, Georges-Louis Leclerc, comte de. *Les Epoques de la nature*, ed. J. Roger. Paris, 1778. Reprint, Paris: Editions du Muséum National d'Histoire Naturelle, 1988.

———. *Histoire naturelle générale et particulière*. 15 vols. Paris: Imprimerie Royale, 1749–67; *Supplément*. 7 vols. Paris: Imprimerie Royale, 1774–89.

———. *Oeuvres philosophiques*, ed. Jean Piveteau. Paris: PUF, 1954.

Burnet, Thomas. *Telluris theoria sacra*. London, 1681. Trans. as *The Sacred Theory of the Earth*. London, 1684.

Caillois, Roger. *Le Mythe de la licorne*. Paris: Fata Morgana, 1991.

Camper, Pieter. *Description anatomique d'un éléphant male*. Published by his son Adrien Gilles Camper, with 20 plates. Paris: Jansen, 1802.

Cardan, Jérôme. *De subtilitate libri XXI*. Nuremberg, 1550.

Cary-Agassiz, Elizabeth. *Louis Agassiz: His Life and Correspondence*. Boston, 1886.

Catelan, Laurens. *Histoire de la nature, chasse, vertus, propriétéz et usage de la lycorne*. Montpellier, 1624.

Cavaillé, Jean-Pierre. *Descartes, la fable du monde*. Paris: Vrin-EHESS, 1991.

Céard, Jean. "La Querelle des géants et la jeunesse du monde." *Medieval and Renaissance Studies* 8, no. 1 (spring 1978): 37–77.

Chambers, Robert. *Vestiges of the Natural History of Creation and Other Evolutionary Writings*, ed. James A. Secord. Chicago: University of Chicago Press, 1994.

Cohen, Claudine. "Approches de la textualité scientifique: pour une histoire culturelle des sciences." *Introduction à l'Histoire des Sciences, Al Madar*, no. 10 (Tunis, 1997): 191–204.

———. "De l'histoire de l'objectivité scientifique à l'histoire des objets de science." In *Des Sciences et des techniques: Un débat, Cahiers des Annales* 45, 149–56, sous la dir. de R. Guesnerie et F. Hartog. Paris: Editions de l'EHESS, 1998.

———. "Leibniz's *Protogaea*: Patronage, Mining and Evidence for a History of the Earth." In *Proof and Persuasion: Essays on Authority, Objectivity and Evidence*, ed. S. Marchand and E. Lundbeck, 125–43. Amsterdam: Brepols Press, 1996.

———. "Un manuscrit inédit de Leibniz sur la nature des 'objets fossiles.'" *Bulletin de la Société Géologique de France* 169, no. 1 (1998): 137–42.

———. "Rhétoriques du discours scientifique." In *La Rhétorique, enjeux de ses résurgences*, ed. J. Gayon and J. Poirier, 131–41. Brussels: Ousia, 1998.

Cohen, Claudine, C. Blanckaert, P. Corsi, and J. L. Fisher, eds. *Le Muséum au premier siècle de son histoire*. Paris: Editions Du Muséum, 1997.

Cohen, Claudine, and J. Neefs, eds. *Science et récit*. Special issue of *Littérature* (March–April 1998).

Conninck, Francis de. *La Traversée des Alpes par Hannibal selon les écrits de Polybe*. Montélimar: Ediculture, 1992.

Corsi, Pietro. *The Age of Lamarck*. Berkeley: University of California Press, 1988.

Cuvier, Georges. "Discours préliminaire." *Recherches sur les ossemens fossiles de quadrupèdes*. Vol. 1. Paris, 1812.

———. *Discours sur les révolutions de la surface du globe, et sur les changemens qu'elles ont produits dans le règne animal*. Paris, 1825.

———. *Essay on the Theory of the Earth, with Geological Illustrations by Professor Jameson*. Edinburgh, 1813.

———. *Recherches sur les ossemens fossiles de quadrupèdes, où l'on rétablit les caractères de plusieurs animaux dont les révolutions du globe ont détruit les espèces*. 4 vols. Paris, 1812; 2nd ed., 1822–24; 3rd ed., 1825.

Czerkas, Sylvia Massey, and Donald F. Glut. *Dinosaurs, Mammoths and Cavemen: The Art of Charles Knight*. New York: Dutton, 1982.

Darwin, Charles. *The Descent of Man and Selection in Relation to Sex*. London: Murray, 1871.

———. *On the Origin of Species by Means of Natural Selection*. 6th ed. London, Murray, 1872.

Daston, Lorraine. "Marvellous Facts and Miraculous Evidence in Early Modern Europe." *Critical Inquiry* 18 (fall 1991): 93–124.

Daston, Lorraine, ed. *Biographies of Scientific Objects*. Chicago: University of Chicago Press, 1999.

Daudin, Henri. *Cuvier et Lamarck: Les Classes zoologiques et l'idée de série animale (1790–1830)*. 2 vols., Paris, 1926–27. Reprint, Paris: Editions des Archives contemporaines, 1983.

———. *De Linné à Lamarck: Méthodes de la classification et idée de série en botanique et en zoologie*. Paris: Félix Alcan, 1926.

Davillé, Lucien. *Leibnitz historien*. Paris, 1909.

Delille, Jacques. *Les Trois règnes de la nature*. Paris, 1801.

Descartes, René. *Oeuvres*, ed. Charles Adam and Paul Tannery. Paris, 1897–1913.

Desmond, Adrian. *Huxley: From Devil's Disciple to Evolution's High Priest*. Cambridge, Mass.: Perseus Books, 1994.

Dewar, Elaine. *Bones: Discovering the First Americans*. Toronto: Random House, 2001.

Dixon, E. James. *Quest for the Origins of the First Americans*. Albuquerque: University of New Mexico Press, 1993.

Dobzhansky, Theodosius. *Genetics and the Origin of Species*. New York: Columbia University Press, 1937.

Efimenko, P. P. *Kostienki I* (in Russian). Moscow-Leningrad: Publications of the Academy of Science of USSR, 1958.

———. *Znatchénijé Jenchtchiny v Orinjiakskoujou Epokhou* (The meaning of woman in the Aurignacian era). 1931.

Eldredge, Niles. "Cladism and Common Sense." In *Phylogenetic Analysis and Paleontology*, ed. J. Cracraft and N. Eldredge. New York: Columbia University Press, 1979.

Eldredge, Niles, and Stephen Jay Gould. "Punctuated Equilibria: An Alternative to Phyletic Gradualism." In *Models in Paleobiology*, ed. Thomas J. M. Shopf. San Francisco: Freeman, Cooper, 1972.

Elster, Jon. *Leibnitz et la formation de l'esprit capitaliste*. Paris: Aubier-Montaigne, 1975.

Febvre, Lucien. "Vers une autre histoire." *Revue de métaphysique et de morale*, nos. 3–4 (1949): 225–48.

Figuier, Louis. *La Terre avant le Déluge*. Paris: Hachette, 1861; 6th ed., 1867. Trans. by Henry Bristow as *The Earth before the Deluge*. London: Chapman & Hall, 1867.

Foucault, Michel. *L'Archéologie du savoir*. Paris: Gallimard, 1970.

———. *Les Mots et les choses*. Paris: Gallimard, 1966.

———. *L'Ordre du discours*. Paris: Gallimard, 1971.

Freud, Sigmund. "Une Difficulté de la psychanalyse" (1917). In *L'Inquiétante etrangeté*. Paris: Gallimard, 1985. Trans. by James Strachey as "A Difficulty in the Path of Psycho-Analysis." *Standard Edition of the Complete Psychological Works of Sigmund Freud*. 24 vols. London: Hogarth Press, 1953–74.

Gaudry, Albert. *Les Enchaînements du monde animal dans les temps géologiques. Mammifère tertiaires*. Paris: Hachette, 1878–90.

————. *Essai de paléontologie philosophique*. Paris: Masson, 1896.

Gesner, Conrad. *Historia animalium*. Frankfurt, 1551.

————. *De rerum fossilium, Lapidum et Gemmarum maxime, figuris et similitudinibus Liber*. Zurich, 1565.

Gillispie, Charles C. *Genesis and Geology: A Study in the Relations of Scientific Thought, Natural Theology, and Social Opinion in Great Britain, 1790–1850*. Cambridge: Harvard University Press, 1951.

Gilmore, Charles W. "Smithsonian Exploration in Alaska in 1907 in Search of Pleistocene Fossil Vertebrates, Second Expedition." Smithsonian Miscellaneous Collections, vol. 51. Washington, D.C., 1908.

Gmelin, Johann Georg. *Voyage en Sibérie contenant la description des mœurs et usages des peuples de ce pays, le cours des rivières considérables, la situation des chaînes de montagnes, des grandes forêts, des mines, avec tous les faits d'histoire naturelle qui sont particuliers à cette contrée*. 2 vols. Trans. from German by Louis de Keralio. Paris, 1767.

Gould, Stephen Jay. "G. G. Simpson, Paleontology and the Modern Synthesis." In *The Evolutionary Synthesis*, ed. E. Mayr and W. Provine. Cambridge: Harvard University Press, 1980.

————. *Wonderful Life: The Burgess Shales and the Nature of History*. New York: Norton, 1989.

Grant, Madison. *The Passing of the Great Race, or The Racial Basis of European History*. Preface by H. F. Osborn. London, 1920.

Guericke, Otto von. *Experimenta nova magdeburgica de vacuo spatio*. Amsterdam, 1672.

Guilaine, Jean, ed. *La Préhistoire d'un continent à l'autre*. Paris: Larousse, 1986.

Habicot, Nicholas. *Antigigantologie, ou Contrediscours de la grandeur des géants*. Paris, 1618.

————. *Gigantostéologie, ou Discours des os d'un géant*. Paris, 1613.

————. *Réponse à un discours apologétic touchant la vérité des géants*. Paris, 1615.

Haeckel, Ernst. *Anthropogenie oder Entwickelungsgeschichte des Menschen*. Leipzig, 1874. Trans. as *The Evolution of Man*. New York: Appleton, 1879.

Hennig, Willi. *Phylogenetic Systematics*. Urbana: University Illinois Press, 1966.

Hunter, William. "Observations on the Bones Commonly Supposed to Be Elephant Bones, which Have Been Found Near the River Ohio in America." *Philosophical Transactions of the Royal Society of London* 58 (1769): 34–45.

Huxley, Julian. *Evolution: The Modern Synthesis*. Princeton: Princeton University Press, 1942.

Huxley, Thomas Henry. *Evidence as to Man's Place in Nature*. London: Williams & Norgate, 1863.

Ides, Evert Ysbrants. *Dreyjährige Reise nach China, von Moscou ab zu Lande durch gross-Ustiga, Sirianan, Permis, Sibirien, Daoum und die grosse Tartarey*. Frankfurt, 1707. Trans. as *Three Years Travels from Moscow Overland to China*. London, 1707.

"International Code of Zoological Nomenclature Adopted by the XVth International Congress of Zoology." London, July 1958.

Jefferson, Thomas. *Notes on the State of Virginia*. 1781. Reprint, Chapel Hill: University of North Carolina Press, 1955.

Kircher, Athanasius. *Mundus subterraneus, in XII libros digestus*. Rome, 1665.

Kotzebue, Otto von. *A Voyage of Discovery into the South Sea and Beering's Straits in the Years 1815–1818*. English trans. (from Russian). London, 1821.

Koyré, Alexandre. *Etudes newtoniennes*. Paris, 1967.

Lamarck, Jean-Baptiste. *Philosophie zoologique*. Paris, 1809.

Lartet, Edouard. "Nouvelles recherches sur la coexistence de l'homme et des grands mammifères fossiles réputés caractéristiques de la dernière période géologique." *Annales des sciences naturelles*, 4th series, 15 (1861).

Lartet, Edouard, and Henry Christy. *Reliquiae Aquitanicae*, ed. Thomas Rupert. London, 1875.

Laurent, Goulven, ed. *Lamarck*. Paris: Editions du CTHS, 1997.

Leibniz, G. W. *Protogaea sive de prima facie Telluris et antiquissimae Historiae vestigiis in ipsis naturae Monumentis Dissertatio ex schedis manuscriptis Viri illustris in lucem edita a Christiano Ludvico Scheidio*. Göttingen, 1749.

———. *Protogaea: The First English Edition*, trans. Claudine Cohen and André Wakefield (forthcoming).

———. *Protogée, ou De la formation et des révolutions du globe*, trans. B. de Saint-Germain. Paris, 1859. New edition, Toulouse: Presses du Mirail, 1993.

Leonardo da Vinci. *Notebooks*, ed. E. McCurdy. New York, 1926.

Leroi-Gourhan, André. *Le Fil du temps*. Paris: Fayard, 1983.

———. *La Préhistoire de l'art occidental*. Paris: Mazenod, 1965. Trans. by Norbert Guterman as *The Art of Prehistoric Man in Western Europe*. London: Thames & Hudson, 1968.

Lindner, Kurt. *La Chasse préhistorique*, trans. Georges Montandon. Paris: Payot, 1941.

London, Jack. *The Call of the Wild*. New York: Macmillan, 1903.

Lubbock, John. *Pre-historic Times*. London: William & Norgate, 1865.

Lyell, Charles. *The Geological Evidences of the Antiquity of Man . . .* London: J. Murray, 1863.

———. *Principles of Geology, Being an Attempt to Explain the Former Changes of the Earth Surface by Reference to Causes Now in Operation*. Vol. 1, London, 1830; vol. 2, 1832; vol. 3, 1833. Reprint, Chicago: University of Chicago Press, 1990–91.

Maillet, Benoît de. *Telliamed, ou Entretiens d'un philosophe indien avec un missionnaire françois sur la diminution de la mer*. The Hague, 1755.

———. *Telliamed, or the World Explained containing Discourses between an Indian Philosopher and a Missionary*. English trans. Baltimore: D. Porter, 1797.

———. *Telliamed: or, Conversations between an Indian Philosopher and a French Missionary on the Diminution of the Sea*, trans. and ed. A. V. Carozzi. Urbana: University of Illinois Press, 1968.

Martin, Paul S., and Richard G. Klein, eds. *Quaternary Extinctions*. Tucson: University of Arizona Press, 1984.

Martin, Paul S., and H. E. Wright Jr., eds. *Pleistocene Extinctions: The Search for a Cause*. New Haven: Yale University Press, 1967.

Maupertuis. *La Vénus physique*. The Hague, 1744.

Mayr, Ernst, *Principles of Systematic Zoology*. New York: McGraw Hill, 1969.

———. *Systematics and the Origin of Species*. New York: Columbia University Press, 1942.

Mély, F. de. *Les Lapidaires de l'antiquité et du moyen âge*. Paris, 1898–1902.

La Mémoire de la terre. Paris: Seuil, 1992.

Michelet, Jules. *Histoire de France: Le Moyen âge*. 1833. Reprint, Paris: Laffont, 1981.

Morello, Nicoletta, ed. *Volcanoes and History: Proceedings of the 20th INHIGEO Symposium*. Genoa: Brigati-Genova, 1998.

Mortillet, Gabriel de. *Le Préhistorique, antiquité de l'homme*. Paris: Reinwald, 1880.

Murray, Tim, ed. *Encyclopedia of Archaeology.* Vol. 1, *The Great Archaeologists.* Santa Barbara, Calif.: ABC-Clio, 1999.

Orbigny, Alcide d'. *Cours élémentaire de paléontologie et de géologie stratigraphiques.* Paris: Masson, 1849–52.

Osborn, Henry F. *The Age of Mammals.* New York: Macmillan, 1910.

Owen, Richard. *Geology and the Inhabitants of the Ancient World.* London, 1854.

———. *A History of British Fossil Mammals and Birds.* London, 1846.

———. *Lectures on Comparative Anatomy and Physiology of the Vertebrates Animals.* London: Longmans, 1843.

Owen, Rev. R. S. *The Life of Richard Owen.* Including T. H. Huxley, "Owen's Position in the History of Anatomical Science." 2 vols. London, 1894.

———. *On the Anatomy of Vertebrates.* 3 vols. London, 1866–68.

———. *On the Archetype and Homologies of the Vertebrate Skeleton.* London, 1848.

———. *On the Nature of Limbs.* London, 1849.

———. "Présidential address, British Association for the Advancement of Science." London, September 22, 1858.

Palissy, Bernard. *Discours admirable des eaux et des fontaines, tant naturelles qu'artificielles, des métaux, des sels et salines, des pierres, des terres, du feu et des émaux.* Paris, 1580.

Pallas, Peter Simon. *Commentaries of the St. Petersburg Academy for 1772.* Vol. 17.

Paré, Ambroise. *Discours de la mummie, de la licorne, des venins et de la peste.* Paris, 1582.

Peintres d'un monde disparu. La Préhistoire vue par des artistes de la fin du XIXe siècle à nos jours (catalogue d'exposition). Musée départemental de préhistoire de Solutré. June 22–October 1, 1990.

Pekarski, P. *Naouk i littératoura v Rossii pri Pietre Viélikom.* Vol. 1. Saint-Petersburg, 1862.

Pewe, Troy L. *Quaternary Geology of Alaska, Geological Survey Professional Paper 835.* Washington, D.C.: U.S. Government Printing Office, 1975.

Piette, Edouard. *L'Art pendant l'âge du renne.* Paris: Masson, 1907.

———. *L'Epoque éburnéenne et les races humaines de la période glyptique.* Saint-Quentin, 1894.

Pliny the Elder. *Histoire naturelle.* Book 7. Reprint, Paris: Les Belles Lettres, 1977.

Poliakov, Ivan S. *Antropologicheskaia poiezdka v tsentral'nuiu i vostotchnuiu Rossijou, Ispolniennaia po porucheniiu Imperatorskoi akademii nauk* (Anthropological travels in central and eastern Russia) 37, no. 1 (Saint-Petersburg: Zapisok Akademii Nauk, 1880).

Prindle, L. M. *The Yukon-Tanana Region of Alaska: Description of the Circle Quadrangle.* Washington, D.C.: U.S. Government Printing Office, U.S. Geological Survey, 1906.

Prokop, Vladimir. *Zdenek Burian a paleontologie.* Prague: Vydal Ustredni Ustav Geologicky, 1990.

Raup, David M. *Extinction: Bad Genes or Bad Luck?* New York: Norton, 1991.

Rigal, Laura. "Peale's Mammoth." In *American Iconology,* edited by David C. Miller, 18–38. New Haven: Yale University Press, 1993.

Riolan, Jean. *Gigantologie, discours sur la grandeur des géants, où il est démontré, que de toute ancienneté les plus grands hommes, & géants, n'ont été plus hauts que ceux de ce temps.* Paris: Adrien Perier, 1618.

———. *Gigantomachie, pour répondre à la gigantostéologie.* Paris, 1613.

Roger, Jacques. "Buffon, Jefferson et l'homme américain." *Bulletins et mémoires de la Société d'anthropologie de Paris*, n.s., 1, nos. 3–4 (1989): 57–66.

———. "Leibnitz et la théorie de la Terre." *Leibnitz, 1646–1716. Aspects de l'homme et de l'œuvre*, 137–44. Paris: Aubier-Montaigne, 1968.

Romer, Alfred Sherwood. "Time Series and Trends in Animal Evolution." In *Genetics, Paleontology and Evolution*, ed. Glenn L. Jepsen, George G. Simpson, and E. Mayr, 103–20. Princeton: Princeton University Press, 1949.

Rosny Aîné, J. H. *La Guerre du feu* (1909) in *Romans préhistoriques*. Paris: Laffont, 1985. Trans. by Harold Talbott as *Quest for Fire*. New York: Ballantine, 1982.

Scheuchzer, Johann Jakob. *Herbarium diluvianum*. Zurich, 1709.

———. *Physica sacra*. Vols. 1–8. Ulm, 1730–35.

Schmerling, Philippe C. *Recherches sur les ossemens fossiles découverts dans les cavernes de la province de Liège*. Liège, 1833.

Schnapper, Antoine. *Le Géant, la licorne, la tulipe*. Paris: Flammarion, 1988.

———. "Persistance des géants." *Annales ESC*, no. 1(January–February 1986): 177–200.

Scilla, Agostino. *La vana speculazione disingannata dal senso, Lettera risponsiva . . . circa i corpi marini che petrificati si trovano in varii luoghi terrestri*. Naples, 1670.

Simpson, George Gaylord. "The Principles of Classification and a Classification of Mammals." *Bulletin of the American Museum of Natural History* 85 (1945): 1–350.

———. *Tempo and Mode in Evolution*. New York: Columbia University Press, 1944.

Solinas, Giovanni. "La *Protogaea* di Leibnitz ai margini della rivoluzione scientifica." In *Saggi sull'illuminismo*, 7–70. Cagliari: Instituto di filosofia, 1973.

Stalin, Joseph. "Marxism and the Problems of Linguistics." *Pravda* (June 20, 1950).

Steno, Nicolaus. *Canis Carchariae Dissectum Caput et dissectus piscis ex Canum genere*. Florence, 1667.

———. *De solido intra solidum naturaliter contento dissertationis Prodromus*. Florence, 1669.

———. *The Prodromus of Nicolaus Steno's Dissertation concerning a Solid Body Enclosed by Process of Nature within a Solid*, trans. J. G. Winter. New York: Macmillan, 1916.

Thenius, Erich. "Phylogenie des Mammalia: Stammesgeschichte der Saügetiere (einschliesslich der Hominiden)." *Handbuch der Zoologie* 8, no. 2 (1969).

Toll, Baron de. "A Geological Description of the Islands of New Siberia: Major Problems in the Research on Polar Regions" (in Russian). *Memoirs of the Saint-Petersburg Imperial Academy of Science*.

Tournal, Paul. "Observations sur les ossements humains et les objets de fabrication humaine confondus avec des ossements de mammifères appartenant à des espèces perdues, par M. Tournal fils, de Narbonne." *Bulletin des sciences naturelles et de géologie* (1831): 250–51.

Des Unicornu fossilis oder gegrabenen Einhorn Welches in der Herrshafft Tonna gefunden worden Berfertiget von dem Collegio Medico in Gotha den 14 febr. 1696.

Velikovetz, L. P. *Johann Georg Gmelin*. Moscow: Nauka, 1990.

Weidenreich, Franz. *Apes, Giants, and Man*. Chicago: University of Chicago Press, 1946.

Wilson, A. F. "The Molecular Basis of Evolution." *Scientific American* 253, no. 4 (October 1985): 164–73.

Witsen, Nicolaas. *Noord en Oost Tartaryen* (in Dutch). Amsterdam, 1692–1705. Reprint, 1785.

Wright, George Frederick. *Asiatic Russia*. New York: McClure, 1903.

Boccaccio, Giovanni, 27, 39; *Genealogies of the Pagan Gods*, 27
Boitard, Félix, 12, 123
Borel, Pierre, 28
Borges, Jorge Luis: "The Garden of the Forking Paths," 217
Bossuet, Jacques-Bénigne, xxix
Boucher de Perthes, Jacques, 81, 127, 148–49, 151, 236
Boulanger, Nicolas-Antoine, 80
Bourguet, Louis, 57, 75
Breuil, Henri, 10
Breyne, Johann, 71–73
British Museum, 93, 220
Brongniart, Alexandre, 116
Brookes, Joshua, xxxii
Buckland, William, 81, 119, 125–26, 127, 192, 232; *Reliquiae Diluvianae*, 125
buffalo, 242
Buffon, Georges-Louis Leclerc de, xxix, 39, 57, 81, 231; on American mastodon, 87, 89–90, 94–96; *Epoques de la Nature (Epochs of Nature)*, 39, 95, 96, 97, 98–100, 145; *Histoire naturelle*, 87, 89; *Histoire naturelle de l'homme*, 97; *Théorie de la terre (Theory of the Earth)*, 81, 97, 98
Buigues, Bernard, 226
Burian, Zdenek, 11, 160–61, 162, 229–30; *The Creator of the Předmost Venus* (1958), 161; *The Mammoth Hunt* (1958), 162; *The Mammoth and Its Young in a Blizzard* (1961), 230; *Mammoths in the Snow* (1961), 11
Burnet, Thomas: *Sacred Theory of the Earth*, 75

Camper, Pieter, 106, 107
Capitan, Louis, 1
catastrophism, xxxii, 112–16, 126–28, 235, 241
Cavazza, Giorgio, 28
cave bear, 150, 241
cave paintings, 5–8, 10–11, 49, 54, 58, 148, 155, 162–63
Céard, Jean, 28

Cesi, Federico, 28
Chaplin, Charlie, 190
Chetverikov, Sergei, 181
Chevalier de Louville, 97
chimera, 59
Christianity, 24–25, 258 n. 2
cladistics, 215–20
Clark, George Rogers, 86
classification, 28–29, 110, 176, 212–13; and cladistics, 215–20; computerization of, 214–20. *See also* evolution; genealogy; phylogenetic trees; scientific names
climate, 97–100, 112, 152, 231, 241–42
cloning, 226–28
Clovis culture, 209, 243
coelacanth, 244
Collinson, Peter, 94–96
colonialism, 128
comic strips, 14, 15–16
Comte, Auguste, 116
Cope, Edward Drinker, 176, 183, 191
Copernican revolution, 145
Coppens, Yves, 66
Correns, Karl Erich, 179
Costa, Philippe, 28
creation science, 146, 235
Croghan, William, 92–93
curiosity cabinets, 2, 28, 41, 64, 77
Cuvier, Georges, xxvii, xxix, xxxii, 2, 36, 40, 81, 91, 128; on American mastodon, 100, 101; *Discours sur les révolutions de la surface du globe*, xxix, 119, 121, 128, 146; on elephant, 105–16, 130, 169; on extinction, 231–32; *Mémoire sur les espèces d'eléphans vivantes et fossils*, xxxi, 105; *Recherches sur les ossemens fossiles de quadrupèdes*, 109, 119
Cyclops, 26

Dall, W. H., 193, 197
Darwin, Charles, xxvii, 127–28, 139–43, 149, 167–69, 239–40; *The Descent of Man*, 149; *On the Origin of*

French Revolution, 105, 123
Frick, Childs, 196

Galen, 33
Garutt, Vadim Evgenievich, 211
Gassandi, 38
Gaudry, Albert, 151, 169–76, 189, 237
Geist, Otto, 196
genealogy, xxix, 139, 167–69. *See also* classification; evolution; phylogenetic trees
genetics, 179–84, 220, 227
Geoffroy Saint-Hilaire, Etienne, 105, 127
geology, 184, 193–94, 231, 235, 250
Gesner, Conrad, 28, 37
giants, 23–40
Gilmore, Charles W., 194–96, 198
glaciation, 118, 193, 232
glossopetrae, 51, 52
Gmelin, Johann Georg, 65–68, 96, 155
Goldfuss, August, 130
gold rush (Alaska), 190–91, 196–97
Goropius [Jan van Gorp], 29, 38; *Gigantomachie*, 29
Goto, Kazufumi, 227–28
Gould, Stephen Jay, 182, 186, 212
gradualism, 184, 188
Grant, Madison, 238
Grayson, Donald, 243
Guericke, Otto von, 44–45, 58
Guettard, Jean, 90
Guizot, François, 124
Guthrie, Dale, 242

Habicot, Nicholas, 31–34; *Gigantostéologie, ou Discours des os d'un géant*, 31
Haeckel, Ernst, 149
Haldane, J. B. S., 181
Haraucourt, Edmond, 12
Haughton, Samuel, 236
Hawkins, Benjamin Waterhouse, 2–4
Hennig, Willy, 215
Herodotus, 27
Herz, Otto F., 200
Hippocrates, 33

hippopotamus, 92, 96
history of the earth: Buffon on, 96–100; and climate, 233; and fossils, 75–81; Leibniz on, 50–51, 53; proof of, 57, 112; and scientific knowledge, 127–28, 249
Homer, 26
Hooke, Robert, 73
Houzé, Jean, 31
Howorth, Henry, 81, 232; *The Mammoth and the Flood*, 81
Hugo, Victor: *Légende des siècles*, 122
Hull, David, 219
humans: evolution of, 149, 183; extinction of, 230, 238; fossils of, 127, 146; and mammoths, 150–63, 208–10, 236, 241–43; origins of, 146, 152, 206; prehistory of, xxxiii, 144–63; tools of, 146–148, 151, 154
Hunter, William, 93
Hunterian Museum (Glasgow), 171
Huxley, Julian: *Evolution: The Modern Synthesis*, 182
Huxley, Thomas Henry, 128, 141, 149, 169

ice age. *See* glaciation
Indians. *See* Native Americans
International Code of Zoological Nomenclature, 212. *See also* scientific names
Isidore of Seville, 27
ivory, xxxiv, 98; in art, 153–55; fossilized, 61, 65–66, 73, 79–80, 197–99

Jacob, François, 181
Jamin, Paul: *La Fuite devant un mammouth* (1906), xxxvi, 1
Jarkov (mammoth), xx, 226–27
Jefferson, Thomas, 85–90, 96; *Notes on the State of Virginia*, 85
Jepsen, Glenn, 182
Johann Friedrich, duke of Hanover, 47

Kircher, Athanasius, 29, 38–39, 42–43, 53; *Mundus subterraneus*, 29, 39, 42, 53

Mayr, Ernst, 181–82
Mazurier, Pierre, 30–31
Mendel, Gregor, 179
Merck, Carl Heinrich, 110
Mertrud, Antoine, 106
Messerschmidt, Daniel Gottlieb, 64, 70–74, 108
Michelet, Jules, 30, 124
Middendorf, Alexander, 66
mining: in Alaska, 190–91; and fossils, 49, 128; Leibniz on, 48–49, 262 n. 24
molecular biology, 220–21
Morgan, Louis, 156
Morgan, Thomas H., 181
Mortillet, Gabriel de, 151
Muller, John Philip, 67
Murchison, Charles, 133
Musée d'Orsay (France), 9
Muséum National d'Histoire Naturelle (Paris), 9, 66, 104, 105, 171, 204
Museum of Natural History (London), 128
myth, xxix, 26, 120–22, 245, 250–51

narrative, xxix, 145–46, 214, 217. See also fiction
narwhal, 42
Native Americans: in Alaska, 197–99; legends of, 87
natural history: as epic, 122; and theology, 64
natural selection, 139, 181–84, 239–46. See also evolution
Nesti, Filippo, 130
Neuville, Henry, 238
New Synthesis, 182–89, 239
Newton, Isaac, 98, 116
Nicholas II, czar of Russia, 203

Ohio animal (Ohio incognitum), 86–88, 90–96, 98, 100, 102
Orbigny, Alcide d', 118, 136, 231
Orosius, Paulus: Historiarum adversus paganos, 30
orthogenesis, 176–79, 183, 238
Osborn, Henry Fairfield, 10, 40, 176–79, 237

Owen, Richard, 28–31, 141, 237; History of British Fossil Mammals and Birds, 130; On the Anatomy of Vertebrates, 128

paleontology: anatomical comparison in, 108–9; controversy in, 127–28; and heroism, 122–23, 251; history of, xxviii, 111–12, 141–43, 249–52; and museums, 171–72; and myth, xxix; reconstruction in, 2, 228; as scientific, 214–15, 249; in United States, 86
Palissy, Bernard, 29, 57, 231, 242
Pallas, Peter Simon, 63, 66–67, 96
Paré, Ambroise, 42
parsimony, 217
Peale, Charles Willson: Disinterment of the Mastodon (1806–8), 102, 104
Pengelly, William, 148
Peter the Great, czar of Russia, 2, 64–66
Pfizenmayer, Eugen W., 199–200, 202–3
phenetics, 214
phylogenetic trees, 167–89, 213–14. See also cladistics; classification; evolution; geneaology
Picard, Casimir, 147
Piestre, Ferdinand, 9
Piette, Edouard, 154
Plato, 129
Pliny, 26, 35
Plutarch, 27
Poliakov, Ivan S., 156
political cartoons, 19
population, 181–82
positivism, 116, 214
Prestwich, Joseph, 148, 150
proboscideans, xxxii, 136–39, 206–8, 211–12; classification of, 215–20; evolution of, 169–89. See also elephant; mammoth; mastodon
punctuated equilibria, 186

Raup, David, 241
Ray, John, 114
Réaumur, René-Antoine, 76

religion, 64, 117–21, 269 n. 37. *See also* Christianity
Riolan, Jean, 31, 33
Rogatchev, Alexis, 222
Roger, Jacques, 100
Rome, 24, 36
Romer, Alfred, 183–84
Rosny Aîné, J. H.: *La Guerre du feu* (1906), 12–13, 144–45, 152; *Vamireh* (1892), 12–13

St. Petersburg Academy of Sciences, 64, 68, 71, 199–200, 203–4, 211
Scheidt, Ludwig, 44
Scheuchzer, Johann Jakob: *Physica sacra*, 75–78
Schmerling, Philippe, 147
scholasticism, 33
scientific experimentation, 56–57
scientific names, xxxii–xxxiii, 101–2, 130, 136, 177–78, 188, 206, 211–12. *See also* classification; phylogenetic trees
Seelander, Nicolaus, 44
Serres, Marcel de, 147
shamanism, 61–63
Shelburne, Lord, 93
Shoshani, Jeheskel, 219
Siberia, 61–73
Simpson, George Gaylord, 92, 182; *Tempo and Mode in Evolution*, 182
Smithsonian Institution, 194
spiritualism, 175
Stalin, Josef, 158
statistics, 214–20
Stenbocq-Fermor, Count, 225
Steno, Nicolaus, 51–53; *De solido intra solidum naturaliter contento dissertationis Prodromus* (1669), 52
synecdoche, 116–17

Tarabykin, Semyon, 199
Tassy, Pascal, 215–16
Tatischev, Vassily Nikitich, 68–70
taxonomy. *See* classification

Taymyr mammoth, 205, 211, 213
Tessier, Henri-Alexandre, 105
Teutobochus (legendary giant), 29–38
Thenius, Erich, 183
theodicy, 175
time, 98–99, 115, 123, 258 n. 2; geological, 167, 235
Tissot, Jean, 30
Tolmachoff, I. P., 237
Tolstukhov, Ivan, 71
tools, prehistoric, 146–148, 151, 154
Tournal, Paul, 147
Trismigestus, Hermes, 41
Tschermak, Erich, 179

unicorn, 41–44, 46, 59–60, 263 n. 51
United States: climate of, 89; geology of, 90; museums in, 102; national identity of, 102–4

Vereshchagin, Nikolai, 205, 220, 243
Verne, Jules, 12
Virgil, 26
vitalism, 236–40
Volosovich, K. A., 66, 204, 225
Voltaire, 57

Weidenreich, Franz, 40
Weismann, August, 182
Wells, H. G.: *Stories of the Stone Age* (1900), 12
whales, 35
Whiston, William, 75
Witsen, Nicolaas, xxxi, 63; *North- and East Tartary*, 63
Wolochowicz, Michael, 72
Woodward, John, 75
Wright, George, 229
Wright, Sewall, 181

Yakuts, 61–62
Ysbrants Ides, Evert, 67

Zamiatnin, S. N., 157